# THE KITE AND THE SNAIL

University Press of Florida

Florida A&M University, Tallahassee
Florida Atlantic University, Boca Raton
Florida Gulf Coast University, Ft. Myers
Florida International University, Miami
Florida State University, Tallahassee
New College of Florida, Sarasota
University of Central Florida, Orlando
University of Florida, Gainesville
University of North Florida, Jacksonville
University of South Florida, Tampa
University of West Florida, Pensacola

# THE KITE AND THE SNAIL

An Endangered Bird, Its Unlikely Prey, and a Story of Hope in a Changing World

*Hilary Flower*

University Press of Florida
Gainesville / Tallahassee / Tampa / Boca Raton
Pensacola / Orlando / Miami / Jacksonville / Ft. Myers / Sarasota

*Cover image:* Snail kite. Photo courtesy of Tim Barker, www.timbarker.com.
*Pages ii, vi, and viii*: Photos by Tim Barker. www.timbarker.com.
*Page xii*: Photo by Hilary Flower.
Cover and text design by Mindy Basinger Hill.



Published in the United States of America

31 30 29 28 27 26 6 5 4 3 2 1

LIBRARY OF CONGRESS CATALOGING-IN-PUBLICATION DATA
*Names*: Flower, Hilary author
*Title*: The kite and the snail : an endangered bird, its unlikely prey, and a story of hope in a changing world / Hilary Flower.
*Description*: Gainesville : University Press of Florida, 2026. | Includes bibliographical references.
*Identifiers*: LCCN 2025053403 (print) | LCCN 2025053404 (ebook) | ISBN 9780813081496 paperback | ISBN 9780813075327 ebook
*Subjects*: LCSH: Everglade kite—Florida | Birds of prey—Florida | Rare birds—Florida | Florida applesnail | Wildlife conservation—Florida—Everglades | BISAC: NATURE / Animals / Birds | NATURE / Environmental Conservation & Protection
*Classification*: LCC QL696.F32 F59 2026 (print) | LCC QL696.F32 (ebook)
LC record available at https://lccn.loc.gov/2025053403
LC ebook record available at https://lccn.loc.gov/2025053404

The University Press of Florida is the scholarly publishing agency for the State University System of Florida, comprising Florida A&M University, Florida Atlantic University, Florida Gulf Coast University, Florida International University, Florida State University, New College of Florida, University of Central Florida, University of Florida, University of North Florida, University of South Florida, and University of West Florida.

University Press of Florida
PO Box 140239
Gainesville, FL 32614
floridapress.org

GPSR EU Authorized Representative: Mare Nostrum Group B.V., Mauritskade 21D, 1091 GC Amsterdam, The Netherlands, gpsr@mare-nostrum.co.uk

*To Tanya*

Systems of people and nature
co-evolved in an adaptive dance.

DR. LANCE GUNDERSON

Adapt. Improvise. Overcome.

ANONYMOUS

# CONTENTS

# FIGURES

# THE KITE AND THE SNAIL

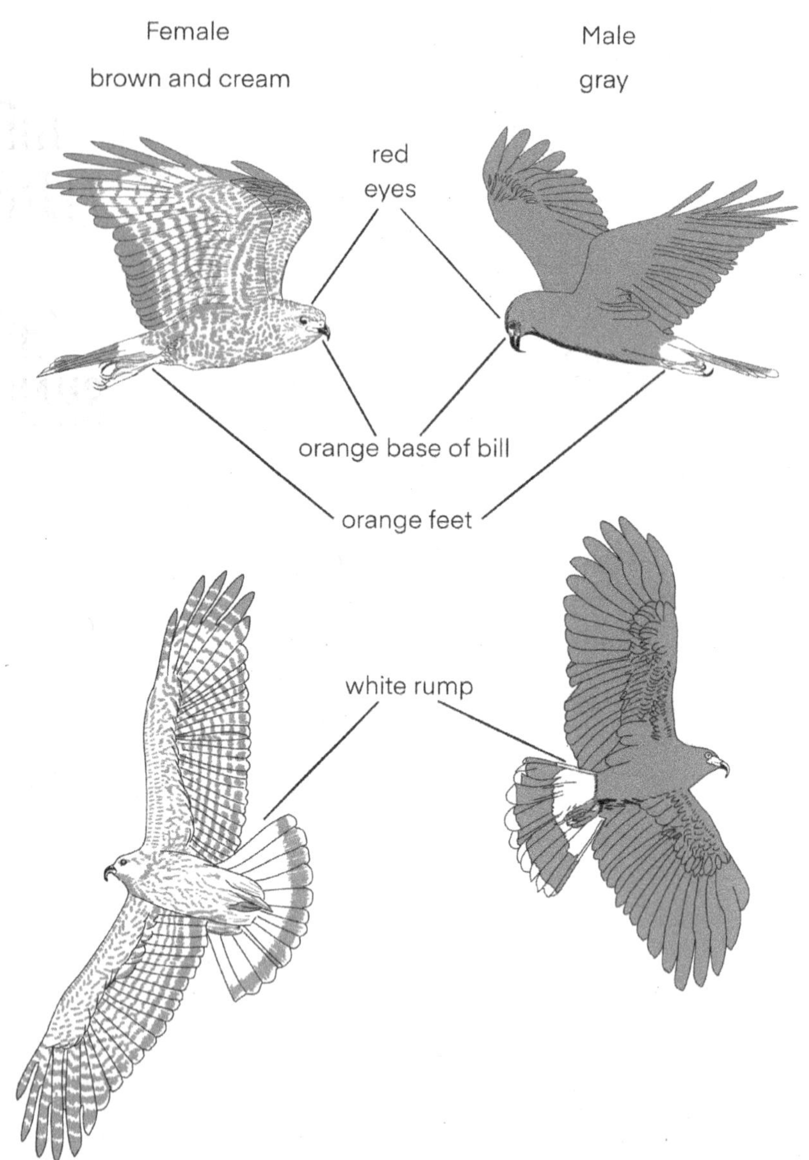


FIGURE 1.1. Snail kite adults have a white rump, orange on their legs and at the base of their bill, and red eyes. *Left:* Females are brown with some cream, including a cream-colored streak near their eyes. *Right:* Males are gray. *Not shown:* Juveniles of both sexes are brown like adult females, with slight differences: more cream on their belly, more orange eye-streak, and legs are more yellow. Illustration by Hilary Flower.

# VANISHING POINT

# 1

"Look, a snail kite!" My ten-year-old Sal pointed to the bird that soared above us and landed on a tree branch not thirty feet away.

I was afraid to hope that I was seeing my first snail kite. The bird resembled a medium-sized hawk, but then I noticed the white band at the base of its tail—a sign that it was indeed a rare snail kite. I felt the thrill of seeing a long-admired celebrity.

The Everglade snail kite is a bird of mysteries. Right off the bat, the name confuses people: Is it a snail? Is it a kite? (And what is a kite?) Kites are a family of raptors known for graceful flight and strong hovering abilities. This one is named for its hyperfocus on eating apple snails in the Everglades. The mystery deepens when you consider that within a decade of my first sighting of the kite, it would lose both the snail and the Everglades, and its graceful flight would take it to very surprising places.

Sal said, "Wait till it turns its head. You'll see the bill is extra curvy."

The bill was so narrow that it was hard to see from the front. When the bird looked to the side, its bill was so looped, I almost laughed. I borrowed Sal's binoculars and was startled to see that its eye was bright red.

Another mystery: Few Floridians know the bird exists, even though it is central to a multibillion-dollar restoration project, and Florida is the only place in the United States where it lives. This raptor is worth knowing, well beyond the bounds of Florida, not just for its quirky beauty, its conservation significance, or its rarity. This bird has something important to tell us. At first, I thought that its message was straightforward, about the plight it shared with countless endangered species. The Everglade snail kite population in Florida was plummeting, but here was an individual kite holding her own. I had brought my three young children and a couple dozen college students to bike on a trail in the Everglades National Park, at the southernmost end of the Florida peninsula. My students all clamored to

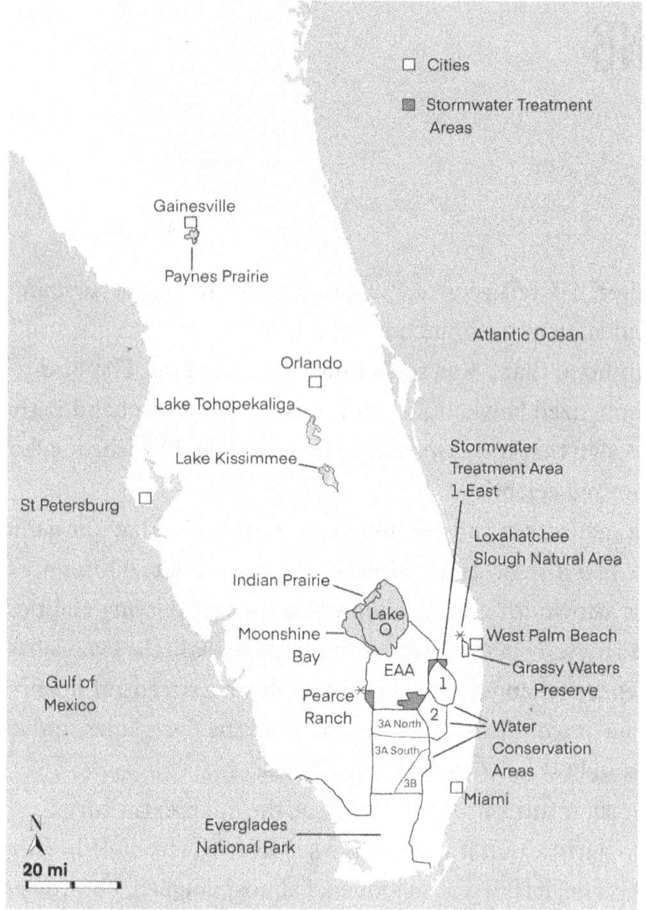


FIGURE 1.2. Featured locations in the snail kite story. Lake Okeechobee (labeled "Lake O") includes the western marshes known as Moonshine Bay and Indian Prairie. Immediately south of Lake Okeechobee is the Everglades Agricultural Area (EAA). South of the EAA are three Water Conservation Areas (WCAs). Between the EAA and the WCAs is a series of Stormwater Treatment Areas (STAs), with the one farthest east being STA-1 East. Immediately south of WCA-3 is the Everglades National Park (ENP). A few miles northeast of WCA-1 are Grassy Waters Preserve and Loxahatchee Slough Natural Area. About sixty miles north of Lake Okeechobee is Lake Kissimmee, and north of that is Lake Tohopekaliga (Lake Toho, for short). Paynes Prairie Preserve State Park is just south of Gainesville in northern Florida. "The Everglades" generally refers to the WCAs and the Everglades National Park. Illustration by Hilary Flower.

get photos of the rare raptor, because they had studied it in wetlands class.

On Everglades trips over the next few years, when I least expected it, I would catch a glimpse of a snail kite, sometimes a brown female, other times a gray male. The thrill of spotting the telltale white bar would last long after it flew out of sight.

In 2013 as we drove out of Everglades National Park, Sal, by then a teenager, called out, "Snail kite!" I looked just in time to see a slate-gray missile with a white tail band fly over my windshield toward the park. That flash of gray and white: I never could have imagined that it would be the last Everglade snail kite I would ever see in the wetland it was named for. After that, whenever I visited the Everglades, I scanned the skies for them, in vain.

The snail kite had always loved the Everglades for its shallow, clear water that made it easy for them to hunt the native apple snails. Although the Everglades had been degraded by a century of human impacts, restoration

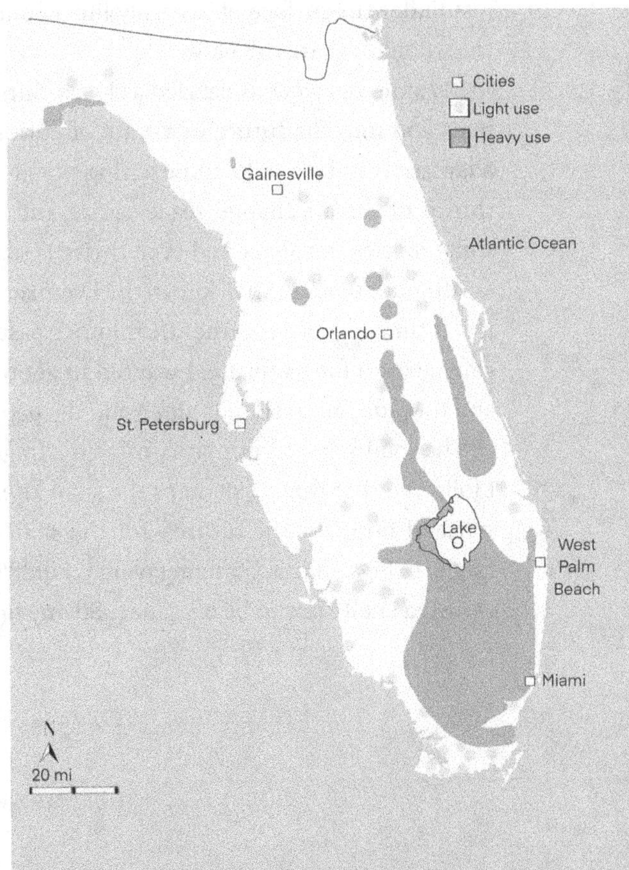


FIGURE 1.3. Historical range of the snail kite in Florida before drainage and development gained momentum around 1900. Snail kites most intensely used South Florida, but they could be found breeding in many wetlands throughout the state. Illustration adapted by Hilary Flower with permission from Paul Sykes Jr. (from Sykes, "The Range of the Snail Kite and Its History" [1984]).

had been under way for over a decade. If the Everglades had become unsuitable for them, I could not detect it; the landscape looked as beautiful as ever to me. Slipping into the liquid world of the Everglades instantly revived my faith in the vitality of the planet's ecosystems, on an emotional if not intellectual level. I made a point of taking students to the Everglades at least once each semester, so they could experience a wild place pulsing with life. Sometimes there were surprises: a tiger-striped swallowtail butterfly, a fuchsia orchid, a brief glimpse of a bobcat, a bear cub, or even a panther.

And then, in 2018, headlines started to pop up: The snail kite population had tripled since 2008! They had "rapidly evolved!" Real evolution in just a few years? How could evolution happen so quickly? What did this mean for their future? And if their numbers were climbing, why was I still not seeing them in the Everglades?

I put these questions aside for later. The delight of the snail kite news shot straight to the sense of ecological dread and despair that often lurked just under the surface of my daily life. I could hardly wait to get into the classroom to share the news.

My students were as dazzled as I was. Surprise is invigorating. Surprise pulls you into the future in a state of curiosity. The snail kites revealed adaptation to be a wild and exciting possibility. As we face the looming threats of climate change, sea level rise, and human impacts, any clue that some species can adapt and even thrive is startling and encouraging.

After not seeing a snail kite in the Everglades for over a decade, I decided it was time to piece together the raptor's past, understand its present, and gain a vision for its future. I wanted to get to know snail kites intimately and to wade into wetlands alongside the people fighting to save them.

The snail kites' cryptic story offered a fragile but real thread of hope. If I followed that thread, perhaps the snail kites could lead me to a sturdier sense of hope for the future—for the snail kites, and for the rest of us fragile beings on this changing planet. I didn't know how I'd get there, but I trusted snail kites to be my guide. So, my first objective was to find one.

# 2

# SEARCHING FOR KITES IN THE WRONG DIRECTION

At 9:30 a.m. on Memorial Day in 2023, I pulled into the shady parking lot at the trailhead for Alachua Sink. Usually, Florida is hot by then, but the air that morning felt crisp, promising a stroll rather than a slog. I had driven two hours north from my home in St. Petersburg to Paynes Prairie Preserve State Park, near Gainesville in northern Florida (see figure 1.2), to meet snail kite researcher Dr. Caroline Poli. I quickly spotted a tall person with a high ponytail and binoculars strapped to her shoulders, the international sign of a serious birder.

Poli had made the local news when she spotted a snail kite while birding with a friend on that trail in 2016. At that time, only seven stray kites had been seen in Paynes Prairie in the prior three decades. In a video recounting the event, she said: "We were rounding the corner, and we both looked to the side at the same time and there was a snail kite sitting on a fence post! Just playing around, extracting a snail. And we both looked, and we were like, 'Oh my God, it's a snail kite!'" She laughed. "We freaked out! And then one bird turned into three, which turned into another bird carrying nesting material . . . And things just skyrocketed from there."

Wow! I was amazed that anyone could be lucky enough to see more than one of this rare species at a time, much less several, much less skyrocketing.

And I was bewildered that this jaw-dropping sight was hundreds of miles north of the Everglades. This raised new questions, but at least it answered the question, "Where did they go when they left the Everglades?" They had moved north. Way north. I could see that, to understand their story, I was going to have to put aside everything I thought I knew about snail kites, the Everglades, and the state I had lived in for twenty-seven years.

I couldn't guess *why* they moved north. And I wondered, Was this a good or bad omen for their future? I was excited to get Poli's take. I was afraid to get my hopes up about seeing a snail kite again after so long.

As Poli and I set off on the wooded path, she told me about the first snail kite she ever saw, a gorgeous gray male in the Everglades back in 2010. Five years later, she started work on a PhD with Dr. Rob Fletcher, who ran the Snail Kite Monitoring Program at the University of Florida in Gainesville. After graduating, she had continued as a postdoctoral researcher in his lab.

As we approached a boardwalk, Poli said, "Snail kites are really sensitive to water."

And there was the water! I was startled at first. The scene before me could not have been more different from an Everglades marsh. Massive oak trees are not an Everglades thing, and yet my view of the lake was framed by the arching limb of a majestic oak tree, dripping scribbles of Spanish moss. Stepping onto the boardwalk felt like entering a Southern Gothic painting. The lake was heavily blanketed with emerald vegetation. I recognized the curly leaves and the blousy lavender blossoms of water hyacinth, a floating aquatic plant I had only seen before in roadside canals.

We followed the boardwalk as it curved around the edges of the lake, leading us to a covered observation deck with built-in benches. We stood overlooking our destination, Alachua Sink. That name holds a clue as to why the snail kites might be there: The lake is a sinkhole. Unique geology formed Paynes Prairie. The Swiss cheese of partially dissolved limestone bedrock, from ancient shallow seas, is overlaid by tens of feet of clay. The clay is a barrier to downward flow of water, allowing large shallow lakes to form. In some places the limestone has collapsed to make a sinkhole, creating a deeper lake. The Alachua Sink is like a kitchen sink complete with a drain; sometimes it gets clogged, and water levels rise. At other times, the water drains out, and the lake levels fall.

Poli told me about how in 2016 hurricanes and storms clogged the drains of Paynes Prairie, and the basin filled up with way more water than usual. Most of the dry land became flooded. We descended a small set of steps from the boardwalk to a grassy trail connecting us to a pond. Poli swept her hands out wide to the horizon: "A few years ago, water was so high that this whole area was flooded. This was not dry land. All the trails here were flooded. You couldn't walk anywhere because of that."

I glanced around, imagining the landscape transformed by open water. Water rippling around trunks of oak trees. Water hyacinth closing over the trail. Something about the change drew snail kites in from far away.

"In 2016," Poli said, "one snail kite showed up, and it was a juvenile, and then more followed. And then in 2018, the snail kites started nesting here. We had like three to four nests. Then in 2019 we had like twenty to fifty nests. So, it took off! And they have been nesting here every year since, as long as the water is high enough."

"Are they nesting here this year?" I asked.

"Yes, yes, the nesting site is really big!"

I looked around us hoping to see a nest, not sure what they looked like. She said that snail kites nest in loose colonies, from a few to hundreds. "They won't nest along these edges here," she clarified. "They'll be out in those willows out there." She placed a hand above her eyes to shade as she looked toward a ragged line of pale green near the horizon. In Florida willows are not grand trees; they are more like tall bushes with skinny branches sticking out at all angles.

"Last year," Poli said, "The nesting in this one specific wetland accounted for 40 to 50 percent of snail kite nesting across their whole range."

I thought I had heard wrong. "The hub for snail kite nesting moved hundreds of miles farther north?" I asked.

"Yes! Paynes Prairie!" she confirmed, nodding with an amused expression. "The snail kites went from not being here at all, and being an incredible discovery, to being something that you can see here any time. And then to this being an important breeding area."

A big splash pulled my attention. An alligator on the bank up ahead had suddenly lurched and plunged into the water.

"There are so many gators in this area," Poli said. "When the water goes down in the dry season, they all come here." She gestured to the dense mat of water hyacinth at the lake edge, noting that even if there were tons of snails amid the vegetation, the snail kites couldn't access them. She said, "The snails become visible underwater when they climb up on the vegetation close to the surface, and the kites plunge-dive down and grab them." She said that there needs to be a window of open water for the snail kites to spot the snail, and room for their wings so they can grab it.

For this reason, if we did see a snail kite that day, it would be at the circuitous boundary between the vegetation mat and the open water. As I traced that curve with my eyes around the lake, I was starting to view the scene from a snail kite's perspective. I noticed that limpkins were wading

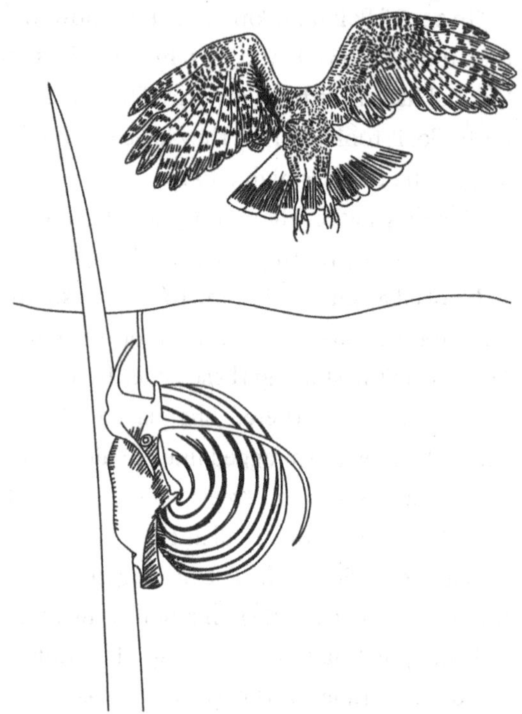

FIGURE 2.1. Apple snails often graze close to the water surface because it allows them to breathe by extending their syphons like snorkels to the air. If the water is clear and the vegetation is well-spaced, snail kites can spot them. Illustration by Hilary Flower.

in the thick water hyacinth. Shaped like a large ibis, limpkins are brown-and-cream wading birds that also love apple snails. Unlike the snail kites, they can find snails by feel, so thick vegetation is not an obstacle as long as there's room for them to wade through it.

Suddenly Poli burst out, "Oh! There's a snail kite right there!" She pointed to a small snag across the water from us, about fifty feet away. "Do you see it?" she asked.

I could feel my heart thudding in my chest as I made out a small brown blob. Even though I could not see it well, I took a deep breath, savoring the moment of once again being in the presence of a snail kite.

Poli was delighted. "There you go! A female." She mused, "This wetland is loaded with them. Even when the water is low like right now, there are so many."

I got my binoculars out and focused in on the kite as she calmly surveyed the marsh around her.

Poli looked around, saying, "These are very dry conditions for her to be foraging well."

I said, "The females are mottled somewhat like a hawk, right?"

"Yeah, like a red-tailed hawk, but more," she said. "And she's also longer and skinnier, and has kind of a little head." She laughed, as if amused at a friend. "I just can't ever get over that."

I said with some hesitation, "The white feathers on its tail look dull, almost tan, a little shabby."

"She's not the most handsome," Poli replied, adding, "She's probably molting. She has probably got some babies somewhere just driving her mad."

I had not expected to get such a powerful sense of the kite before me as an individual, with a life story and her own problems. I looked more closely at the irregular pattern of brown and cream on her chest. I realized that each female kite must have her own pattern, like a fingerprint.

Poli pointed to her ear, saying, "Listen. You can hear a snail kite yelling."

I had assumed it would sound majestic like a red-tailed hawk. It was more like a croak, rough and bumpy, like scraping on a metal washboard. I later saw the kite's cry described as "a series of quick, short notes that sound like a wooden dowel turning in a tight hole."

Poli noted that snail kites are very social, especially for a bird in the hawk family. Their scientific name is *Rostrhamus* (hooked beak) *sociabilis* (sociable).

I asked her why snail kites had come to Paynes Prairie, and why they had left the Everglades. She said that the short answer was snails. Snail kites only eat apple snails, and Florida has only one native species of apple snail, whose common name is the Florida apple snail. In the early 2000s, the native snail started declining. Around that time a non-native snail was becoming abundant in the central part of the state, and the snail kites followed them there. Then, when the storms of 2016 suddenly expanded the wetlands of Paynes Prairie, the non-native snail took off, and the snail kites homed in.

This explanation raised even more questions; I realized that part of my quest to understand the Everglade snail kite story was going to be getting to the bottom of the mysteries embedded in its name: the snails and the Everglades. What was going wrong for the native snail in the Everglades?

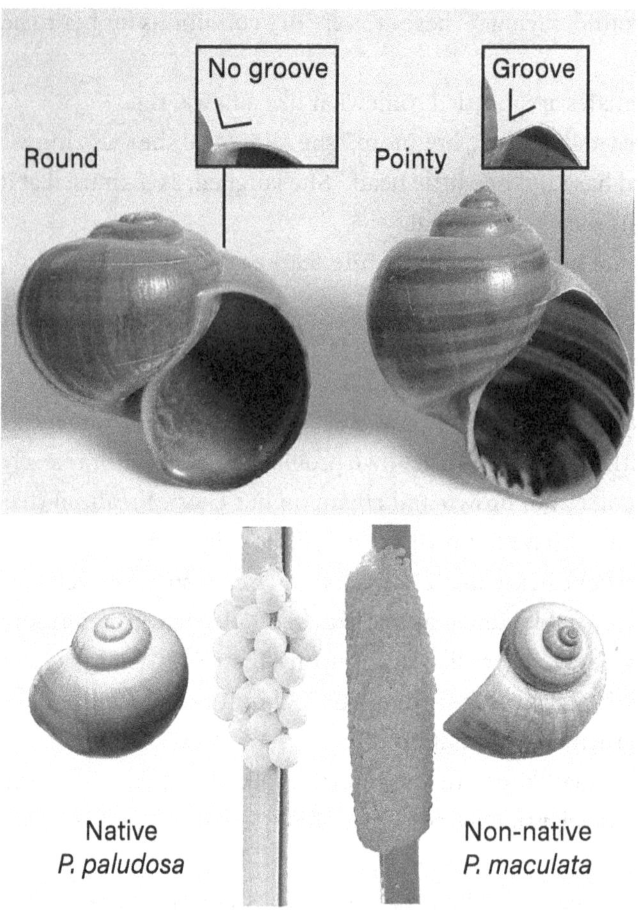


FIGURE 2.2. Two snail species are central to the snail kites' story in Florida: Florida's only native apple snail, *Pomacea paludosa* (*left*), and Florida's only non-native apple snail that has been designated invasive, *Pomacea maculata* (*right*); key distinguishing features are the shell of the invasive snail having a pointier top and a deep groove in its spiral. *Bottom:* Egg clutches for the native snail (*left*) with white eggs and invasive snail (*right*) with pink eggs, as shown on blades of saw grass to demonstrate the relative sizes of both eggs and clutches to one another. Illustration by Hilary Flower.

Why hadn't the non-native snail taken off in the Everglades? What role would these two snails, and these two parts of the state, north and south, play in the snail kites' future?

Poli pointed me to snail experts who could help solve those puzzles. For that day, I was thrilled to be near a snail kite again, to see them in their very different new home, to get to know a real snail kite expert, and to have some initial answers as to how they got there.

A kite at the far side of the lake began to fly our way. As it approached us, it started to get larger than I thought possible. Snail kites have a wingspan of more than three feet. I couldn't believe my eyes when it flew over to perch in the tree right next to the observation deck where we were standing. We could almost reach out and touch it. We went up to the railing by it and marveled at its curved bill. The tail band on this one was brilliant white. The shifting dapples of sunlight played on her mottled chest.

Poli leaned over the railing, exclaiming, "Look there's four right here!"

I followed where she was pointing as snail kites began to swirl around us.

She said, "See: one, two . . . three . . . four, there's a fifth one right there. . . . Here's number six!"

And it just skyrocketed from there.

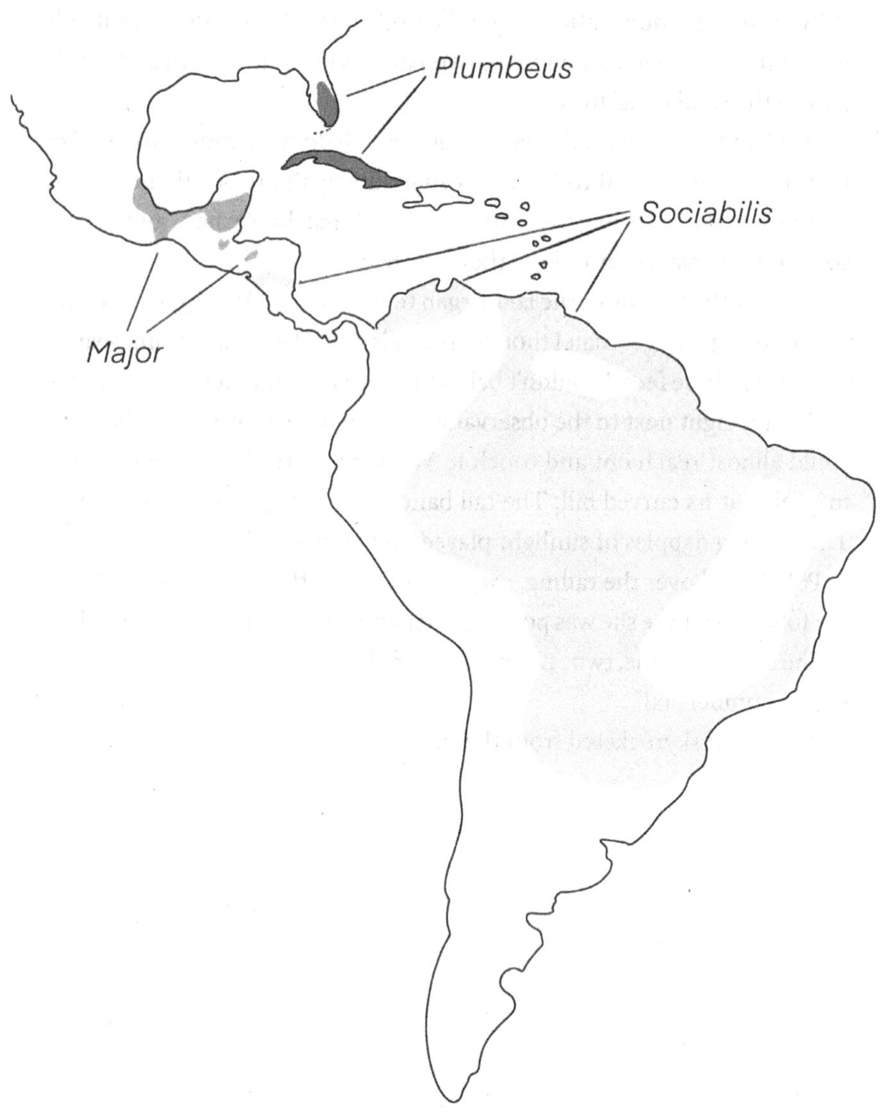


FIGURE 3.1. There are three subspecies of snail kite: *R. sociabilis plumbeus* lives in Florida and Cuba; *R. sociabilis major* lives in Mexico and Central America; and *R. sociabilis sociabilis* lives in South America. Illustration adapted by Hilary Flower from Haas et al., "Genetic Divergence Among Snail Kite Subspecies" (2009).

# 3 THE EXTREME SPECIALIST RAPTOR

Of all the things Dr. Caroline Poli told me on our walk at Paynes Prairie, one thing stuck with me the most. I had asked her, "Since there are snail kites in South America that are doing well, why would it matter if the Everglade snail kite went extinct in Florida?"

I knew the snail kite range in the US was limited to Florida, and that globally there are three subspecies. Our subspecies, the Everglade snail kite, bears the scientific name *Rostrhamus sociabilis plumbeus,* where "plumb" refers to the males' lead-gray coloring, like old plumbing pipes. The Florida subspecies also lives in Cuba, although they don't seem to go back and forth. Unfortunately, almost no research has been published on snail kites in Cuba, so we have no population estimates, and their conservation status is unclear. There are two other snail kite subspecies: one in central America, and another in South America, which hosts the species' largest population. The subspecies are visually identical, but they can be distinguished genetically.

I was glad that snail kites did not appear to be in danger of extinction on the planet. As I set about my quest to understand their plight in Florida, I wanted to know what the stakes were. Why does our subspecies in Florida matter?

Poli said that, first of all, she'd be really mad if they were gone, if she went out to a place like Paynes Prairie and there were no snail kites. I realized that I had experienced that reality in the Everglades.

I wondered how prevalent that reaction would be. Most of the people I saw at Paynes Prairie seemed excited about the gators and oblivious to the snail kites. Then I remembered that bird-watching and wildlife viewing in general are thought to bring more than $5 billion annually to Florida's economy. Later that morning a man dressed in camo came to the observation deck and took photos of the snail kites using a camera with an extremely long lens. I met an older couple who had come from Michigan to see as many birds as they could. The snail kite was one of the birds they had hoped to add to their "life list," an ongoing tally some birders keep of the species they've seen.

Then Poli raised the discussion above the personal and human, and I ended up spending a lot of time unpacking the meaning of her words: "The snail kite holds a particular place in biodiversity. The snail kite is unique. It's a specialist predator. There are very few of those, much less *raptor* specialists. And as specialists go, they are one of the most extreme. Without the snail kite, this whole approach to being an animal in the US would be gone."

Whoa. On my drive home that day, I turned her statements over in my mind. I knew that to specialize on snails, kites had unique adaptations and behaviors, but I hadn't thought of their uniqueness as an added reason to save them.

The snail kite's body and its behavior are tailor-made for snail hunting. The mystery of the snail kite, its superpower and its kryptonite, its past and its future, all revolve around its bold and risky strategy of eating only apple snails. The snail, I realized, is the key to all the ways in which the kite's body and behavior deviate from a hawk's.

Despite the dazzling display of snail kites on my walk with Poli, I had not seen one catch a snail. If rapid evolution was one of the mysteries I needed to unravel, a starting place was to fully understand how snail kites had evolved to hunt snails in the first place.

I planned a trip to return to the observation deck at Paynes Prairie to try again to witness this behavior. In the meantime, I did my homework, reading everything I could on snail kite adaptations.

My first priority was to make sure I understood the terms that Poli had used. Specialists are animals that eat only one thing, or mostly one thing, and their bodies adapt accordingly. Koalas eat only eucalyptus leaves; their digestive system is adapted to subsist on a plant that is fibrous, low in nutrition, and highly poisonous to most other animals. The more humans cleared eucalyptus forests, the more endangered koalas became. Pandas eat bamboo almost exclusively; they have evolved special wrists for grasping it. As bamboo habitat became scarce, they, too, became endangered.

Specialist herbivores are one thing; specialist predators are even more rare. They need habitats that support their prey. Black-footed ferrets specialize in hunting prairie dogs, and they were pushed to the brink of extinction as prairie dog habitat disappeared.

Evolution had allowed these species to capitalize on the most abundant resources in each of their environments. And we have pulled the rug out from under them.

Poli had noted that it's especially rare for raptors to be specialists. Raptors are birds of prey; they hunt other animals for their meals. Many raptors specialize as to the *category* of prey they hunt, but not the species. For instance, the osprey is all about fish, and it will eat many species of fish. My ornithologist friend Dr. Beth Forys said that she once saw a squirrel go up into an osprey nest and the osprey seemed to have no idea what to do about it. She said that Cooper's hawks specialize in hunting other birds like doves.

The other animals that eat adult apple snails, like limpkins and redear sunfish, also eat other shellfish, such as crabs, clams, and other snails. The snail kite stands out as an extreme specialist. If apple snails are available, that's all the snail kites will eat. One study reported apple snails comprising 99.4 percent of the kites' diet.

I wanted to learn everything I could about snail kites' adaptations to hunting snails, so I reached out to Dr. Steve Beissinger, who recently retired from the University of California at Berkeley. Beissinger did breakthrough field studies on snail kites in the 1980s. He went on to become famous not only in the world of kites but also of California birds and mammals. I was thrilled when his face appeared on my screen for our first virtual interview. His big smile was framed by a trim salt-and-pepper goatee. He appeared to be in a sunny study with many bookcases.

I asked him, Why snails? Beissinger pointed out that one benefit of specializing on snails, instead of mammals or other birds, is that snails are nonthreatening. Most prey evade predators through maneuverability, struggling, and sometimes fighting back. With a wry smile he said, "Snails rank low on these qualities. Not much maneuverability or strength is needed to catch and eat them. Snails seem relatively tame compared to catching a bird in midair, for example."

It was a trade-off: By evolving to target such a specific prey item, snail kites became worse at hunting other animals. Beissinger explained, "I think it's not easy for them to capture and kill fish, birds, and small mammals with their long decurved bill and wimpy talons. But when they are hungry enough, they do so." They've been known to eat crabs, crayfish, fish, and mice. If they are desperate enough, they turn to turtles, but it can take them hours, and they don't get much.

Their curvy bill is their most famous adaptation, but they use a lot of other adaptations to get the snail in the first place. With Beissinger's help, I strove to identify all the challenges of hunting apple snails, and all the ways in which snail kites adapted to overcome them. Then I returned to Paynes Prairie to witness it all in action. When I arrived, the bank of forest on the far side of the lake took on a golden haze, backlit by the sun rising behind it. I was conscious that it is always risky to go out into nature with an agenda. But I couldn't help it. I was on a mission to see the snail kites' signature move, what Beissinger referred to as "hover and pluck."

Before long, a snail kite emerged from the oaks across the water and began to patrol the edge of the lake where the mat of water hyacinth gave way to open water. The distant bird made a small, dark shape interrupted by the white at the base of its tail. The tops of its wings intermittently flashed with golden sunlight. As the kite became ever more distant, the shock of white became its most visible part. I watched that shrinking white bar lift and glide.

This snail kite was up against the first challenge the raptors must overcome. Apple snails are notoriously hard to spot, even for experts. They only emerge into air to lay their eggs on stalks above the water surface, usually under the cover of night. Apple snails are "amphibious": They have gills and lungs, so they can breathe underwater like a fish. And when they want to breathe air, they have a breathing tube like a snorkel. They have to get within one to two inches of the surface to use their snorkels: That's what allows snail kites to see them.

Apple snails have only one evasive measure: quick release. When startled, they drop to the ground. To catch them near the surface, a predator must grab them unawares from above.

Spotting a small object underwater is not easy, but snail kites have eagle eyes. All raptors devote a third of their skull cavity to their eyeballs. This means there is no room in their skull for muscles to control eye movement; that's why they must turn their head if they want to look to the side. Hawks can spot a mouse in a prairie one hundred feet away. Snail kites can zero in on snails at a distance of thirty feet or more.

Eventually the snail kite's circuit brought it to my side of the lake. Watching it fly by, I was a bit unnerved to see that it did not look ahead, only down. I almost wanted to say, "Hey! Look where you're going!" The top of its head pressed forward, its bill pointed down, and its eyes seemed to shoot lasers into the water below.

Another trick that helps many predators spot their prey is that they tune into a specific visual pattern, known as a "search image." This allows them to quickly scan a large area for the desired item, like the search function on a computer. The trade-off is that using a search image can lead to tunnel vision for that one prey item, making it harder to spot other prey. I had asked Beissinger if snail kites have this visual trick, and he said it was plausible, but it hasn't been studied.

To complement their acute vision, snail kites have some behavioral adaptations: They hunt in still water because ripples obscure their view. Snail kites will forage on the leeward side of an island, even if there are more snails on the windy side. Hunting in marshes mitigates wind problems because vegetation absorbs breezes, allowing the water surface to remain glassy.

Snail kite wings also help them spot snails because their shape maximizes the birds' ability to hover. Birds in the kite family tend to be small and light, with a small head and a small bill. Those endowed with narrower wings are built for speed, and wider wings are better for hovering and slow gliding. Snail kite wings fall on the wide end of the spectrum. When foraging, snail kites fly into the wind, which makes it easier to hover. The net effect is that snail kites can fly more slowly than just about any raptor, staying aloft at speeds that would cause other birds to drop like rocks.

When they spot a snail, they drop down and punch their feet through the water to grab it. Now, the next set of challenges arises. Slippery, wet, and round objects are hard to carry. Snail kites have extra-long legs, toes, and claws, also known as talons. It's as if the legs and feet of an ordinary hawk were stretched to maximize snail-holding. Their talons look like steely

FIGURE 3.2. Snail kite foraging for a snail. Illustration by Hilary Flower.

FIGURE 3.3. The snail kite's bill has evolved to curve just right to reach into the snail's shell and sever the columellar muscle that tethers the snail's body to its shell. Illustration by Hilary Flower.

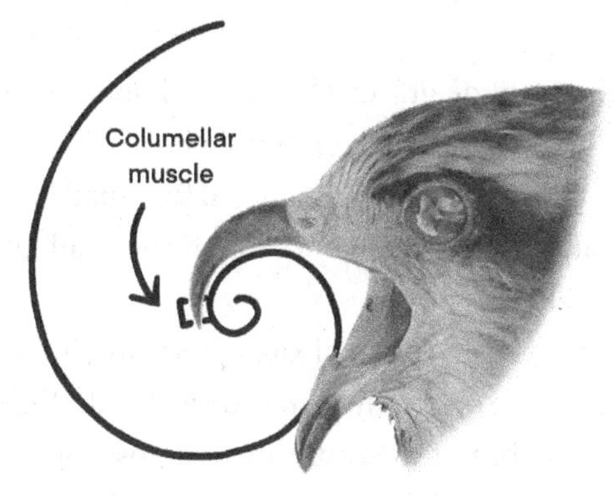


daggers, but they are more like catcher's mitts. Most raptors would break the skin if they landed on your arm, but not snail kites. Snail kites make good use of the extra length; they can reach six inches below the surface without even getting their feathers wet. Their long toes and talons wrap around the snail, yanking it from the vegetation and securing it for flight.

The hard snail shell locks out many predators, most of whom would have to swallow it whole and hope for the best. The redear sunfish, also known as the shell cracker, is specially adapted to eating shellfish. They have grinding plates in their throats where they can crush shells, and they can spit out shell fragments before swallowing the meat. Instead, snail kites have evolved to separate the snail meat from the shell before ingesting it.

But doing that isn't easy. Apple snails can withdraw into their shell and slam the door behind them. Their "door" is a brown oval-shaped disk made of fingernail-like material that perfectly closes just inside the shell opening. It's called an operculum, Latin for "little lid." The operculum keeps predators out, and during dry periods, it keeps moisture in.

Also, the snail's body is strongly attached to the inside of the shell with the columellar muscle. This is where the snail kites' specialized bill comes into play. First, it is like a letter opener: thin, sharp, and strong. The snail kite typically perches and holds the snail with one foot while it inserts its bill between the operculum and the shell. Once in, they can use their bill to pry back the operculum, detach it from the snail's flesh, and drop or fling it. Next, the bill curves just right to reach around the inner curve of the snail shell to sever the attaching muscle with a sharp poke or two. Finally, the coup de grâce: The pointed hook skewers the snail's body and yanks the coils of flesh from the shell.

For over an hour I watched two snail kites take turns patrolling the lake. I ate a tangerine, wondering if the snail kites were ever going to get their breakfast.

Suddenly, the snail kite I was watching started to descend with its wings raised up vertically alongside its head. With barely a splash, its feet dashed in and out of the water in a microsecond. It vigorously pumped its wings to lift up and over to a nearby tree branch, its prize clutched in its talons. It had caught a snail from a little patch of open water within the large mat of water hyacinth.

I watched it on its perch as it fussed with the snail, arranging with its talons and its bill. It pecked, tugged, flung something (the operculum, I guessed), and started taking bites and swallowing. It was done in a minute, and it spent a good while wiping the side of its bill off on the branch. It reminded me of someone dabbing the corners of their mouth with a linen napkin.

I had finally seen a snail kite hunt! This is what the birds are named for. This is what they were born to do, for generations untold.

I lingered to witness it again and again, until the behavior and its connection to everything in the wetland around it sank into me at a visceral level. At one point, I was watching a female snail kite perched on a willow branch on the other side of a small canal. I did not realize she was hunting until she abruptly shot over to grab a snail from the water, which she brought back to her perch. I had read about this strategy. Hunting from a perch, often with the sun at their backs, is known as "still-hunting," whereas hunting by flying over the marsh is known as "course hunting." I could not believe she had spotted a snail from fifteen feet away, especially as she seemed so nonchalant. As other kites flew past her, she called out. I wondered if it was a friendly greeting, since Poli had noted that they are sociable.

I drove home that day with images playing in my mind's eye of snail kites flying, hovering, grasping, carrying, perching, and extracting. I felt like I had cracked the code for what makes them unique, and what would be at stake if they went extinct in Florida. Over thousands of years, they had evolved an impressive array of physical and behavioral adaptations that made them supremely suited to hunt snails. I loved to think that the inner curve of the snail's shell had produced a matching curve in the kite's bill.

Lakes are not a big part of the Everglades. It got me wondering what it would be like to witness them foraging in the historical, predrainage (pre-1900) Everglades. To my surprise, I was invited out to do just that, or at least, the closest modern equivalent.

# 4 THINKING LIKE A KITE

In the predawn darkness, Gina Kent laid out her plan to trap the kite. Kent is the senior conservation scientist for a nonprofit organization called Avian Research and Conservation Institute (ARCI), where she has worked since 2000. Dr. Ken Meyer cofounded ARCI in 1997, and he directs it. ARCI studies imperiled birds to inform management and conservation efforts. Kent was trying to trap this kite in part to test her for exposure to an environmental toxin known as methylmercury. By putting a lightweight GPS tracker on her, they can also find her future nest, so they can (safely) test her future eggs, and monitor the nest for hatching and fledging success.

I had been out there with Kent for two days already. I certainly hoped to catch a kite with her someday, but my number-one goal for the trip was to learn more about these elusive birds.

We were standing on a levee in Loxahatchee Slough Natural Area in northeastern Palm Beach County, about forty miles east of Lake Okeechobee (for location, see figure 1.2). Before us was a canal, behind us a swamp. Next to a post in the canal edge, Kent had laid two large floating trays, with tethered snails and a lattice of fishing line to catch the kite's feet. The previous day she had seen a female snail kite perch there for a few hours, intermittently grabbing a snail from the shallow water. The canal edge had very few posts, so Kent was confident that the same kite would return.

In the cypress swamp behind us was a "communal roost" where 146 snail kites, and even more wading birds, had passed the night. In Florida, roosts are typically a stand of trees surrounded by water deep enough to keep raccoons out. A great blue heron gave a loud, raucous cry, and soon after, we saw its regal silhouette as it flew over the canal. A long stream of wading birds followed suit. I could dimly make out the shapes of little blue herons, great egrets, white ibises, and more.

Among them were snail kites. I watched their silhouettes reflected in the canal. Kite after kite after kite. A flowing river of flapping wings. The sky broke out in pink, yellow, and lavender near the horizon, and soon the sun was visible through the trees across the canal. I watched in awe as more than a hundred snail kites flew over the canal.

None of the snail kites seemed to give the post a second look. Then one descended and perched. Kent had instructed me not to make eye contact with the snail kites or turn my head to follow their flight. If they catch you watching them, they could read you as a predator.

Kent pulled up her binoculars and verified that it was a mature female, based on the bright orange of its legs, its mostly brown belly, and the cream-colored streak near its eye. She said that streak would be more orange if it were a juvenile, its legs more yellow.

The perched snail kite cackled when others of her kind flew past her over the canal. Kent said the kite was making it known that this foraging spot was hers. I realized that must be what the perched kite at Paynes Prairie had been doing, too.

"Do you hear that?" Kent asked, as a keening sound came from the cypress trees across the canal. "That's called reaping," she said. "It's a young kite calling to its parents to feed it. It could be this female's offspring." She explained that snail kite parents continue to bring snails to their fledglings for weeks after they leave the nest. Learning to hunt snails is not easy.

Kent studied the kite's feathers through binoculars. "Yep," she said, "she's molting alright." She explained that with a young kite, the white edges at the tips of feathers line up and form a continuous curve. But within a year, the feathers become worse for wear. Birds replace old feathers over time throughout the year, gradually and often symmetrically. They start this at their second year. Kent noted that this kite's feathers were uneven, the telltale sign of molting. That was good because she only wanted snail kites at least two years old.

The kite spent a lot of time on her perch looking down toward the traps and the surrounding water. She flew off to grab a snail from just beyond the traps. She brought her breakfast back to her perch. Kent said that snail kites tend to eat about a snail an hour, and she set about waiting. Eventually, though, she decided to try elsewhere. Snail kites are individuals, after all, and this one was extremely cautious. The next time she flew off, Kent took the opportunity to retrieve the traps.

We drove along the levee until we came to a culvert. Kent climbed down to the water's edge and waded among the rocks, looking for snails. I did the same, and I was proud to spot one—only to find that it was empty. Kent had the eyes of a kite: She quickly found three. Back at the truck, she dropped them into a bucket with water and vegetation.

As we drove to find a good marsh for setting up traps, each of us spoke up when we spotted a "brown blob," which were silhouettes of birds perched in distant trees. Often it was a limpkin. Once it was a peregrine falcon. We also called out if we spotted the flash of white indicative of a snail kite in flight.

Soon we stood on a levee overlooking a wetland with bright-green lily pads and sparse, scraggly cypress trees. Kent had her binoculars trained on a female snail kite in a cypress tree about a hundred yards from the levee. She noted with a big grin that she had seen it in the same spot the day before. Kent was confident that this was the kite's "plucking perch," which made it a perfect spot to put the traps.

We got the four traps out of the back of the truck and fished snails out of the bucket. Some of them had little clips glued to their backs, which we could attach to a little leash in the bottom of the tray. For good measure, Kent tossed in one of the snails she had found that morning, even though it didn't have a tether.

When the bird flew off, Kent and I slipped kayaks into the water and paddled over. Whenever I think back to that wetland, my mind's eye fills with blue and green. Blue sky reflected between emerald circles. Our paddles slid between lily pads with yellow-centered white flowers, a native plant known as fragrant water lily. This resembled a marsh in the historical Everglades, quite a contrast to the deep, water-hyacinth-rimmed lakes at Paynes Prairie.

One thing I've always loved about fieldwork is how it takes you into places you never would be otherwise. This preserve was maintained by Palm Beach County's Environmental Resource Management (ERM). Kent told me the ERM had done a lot to maintain and protect this important part of the historical Everglades watershed. The public was only allowed to walk or bike on the levees, but the ERM folks had given Kent a key to the big yellow gates so that she could drive her ARCI field truck along the levees. They'd even loaned her a kayak for the day so we would have two.

As I took in the beautiful marsh all around me, I became one with it in unexpected ways. Garlands of submerged vegetation got caught in my paddle and came down on me with each stroke. I was sporting quite a bit of soggy foliage by the time I figured out how to angle the paddle better.

Ahead of me, Kent was at the cypress tree, looking it up and down. She called me over with a gleeful expression. "Look at that tree," she said. "What do you notice?"

"Well, that one branch looks kind of horizontal," I said, "like it might be good to perch on?"

She replied, "Look how dark brown that part is! That's a lot of snail goo. Many snails have been extracted there. And look here," she pointed with an appraising smile at what looked like white paint spattered on the lily pads below the branch. "Bird poop! Kites have used this perch a lot."

She set up one trap there and another at a lone cypress tree a ways off, where she had seen another kite perch the day before.

We made our way behind a patch of tall saw grass a couple hundred feet from the kite. We had a line of sight to each of the traps. About a half hour later, a female kite landed on the perch. Even more exciting, our binoculars revealed her intently staring down at one of the traps. She moved her body a bit forward and back and slightly tipped her head as she gazed at the snails in the trap. Kent said this behavior, called tracking, was what we wanted to see; it improves a raptor's view of their prey. Our kite was considering going for one of our snails.

Suddenly, the kite flew down and hovered above the trap, legs extended, and Kent held her paddle in both hands ready to go. She had been ready to bolt because once a snail kite gets caught, the clock is ticking. Kent is committed to minimizing distress to the kite, so every second counts.

But the kite did not take a snail. Instead, she flew back up to her perch. She turned to face the opposite direction and fluffed her feathers.

I asked, "Is she uneasy? Maybe she's concerned that we are watching her?"

"No," Kent said with confidence. She explained that when birds fluff out their feathers it's called rousing, and it indicates that a bird is comfortable, not stressed. Same with preening.

FIGURE 4.1. Gina Kent prepares a tray of snails to catch a snail kite, below a cypress tree in Loxahatchee Slough Natural Area, as part of her work with Avian Research and Conservation Institute (ARCI). Photo by Hilary Flower.

The kite turned to face the trap and began tracking again. She was clearly tempted. She stared down for long stretches, then turned away to rouse and preen. At length, she turned again and stared down at the trap. All at once, the kite flew down to the trap, grabbed a snail, and flew back up to her perch to consume it. It must have been the untethered snail, and somehow the lattice of fishing line had not caught her foot.

I thought Kent would be disappointed, but she was smiling. "This is good," she said. "She was rewarded for taking a snail from the trap. That makes it more likely that she'll try again." Kent is proficient at the long game: She can sit in silence indefinitely, poised to spring. She was used to an open-ended kind of fieldwork, relying on the behavior of a creature with a mind of its own. Success might happen all at once, or not at all.

A while later, the kite hovered over the trap again, and in a flash, she was pulling at a tethered snail. Then she let it go and flew back up to her perch. The fact that her feet had escaped capture both times was pure luck: good luck for her, bad luck for us. Eventually, she flew off for good.

Our attention turned to a male snail kite perched on the lone cypress tree. As we eagerly watched, he flew down to hover over a snail in the trap, his wings high and his feet an inch above the trap. So close! But then he flew back up to his perch. He stared down at the trap, cackling loudly. Another kite not too far away responded. This sequence repeated about ten times: hover, retreat, cackle, and a response from the other kite. But he never took a snail.

We removed the traps and tried another spot, and our luck was no better. As the sun started to approach the horizon, we had to pack up. As Kent drove along the levee, I thought about how she would have to make the four-hour drive back to her home in Gainesville and then plan another whole trip. Yet Kent's spirits were high. She had identified the most popular kite perches for her next trip. She was ready to work with snail kites on their terms and on their timetable.

My spirits were high, too. I had witnessed more than a hundred snail kites emerge from a huge communal roost at dawn. I had learned the value of a perch for still-hunting, and how kites vocalize to warn others off. I had seen the telltale signs of a plucking perch, and watched snail kites track, preen, rouse, hover, grab snails and eat them. I could still hear the poignant reaping of the unseen fledgling across the canal. I now knew how to spot molting, and how it helps to age a bird. I had learned how to tell if a kite is calm, and how to avoid being perceived as a predator. My time out there with Kent and the kites had given me a little portal into what it is to be a snail kite in a place that looked like the Everglades.

"There's a kite," I said, out of habit, when I saw one perched on a cypress tree we were passing. I didn't think Kent would take the time to stop, given the long drive ahead of her.

But she stopped the truck and backed up a few inches. We raised our binoculars. His feathers were a rich gray, almost blue. His feet were a shock of orange against black talons. The base of his bill was the same fiery orange against the curved black tip. When he turned his head, we could see the red of his eye.

We watched him until he flew off to a distant cypress dome.

# BACK TO THE BEGINNING

# 5

Although the blue and green marsh I had been in with Kent resembled the historical Everglades, the snails there were not the native ones. Somehow, the relationship between the native apple snail and the Everglades had been broken. I wondered if it could be repaired. To begin to answer that question, I had to first understand how those three had evolved together: the native snail, the snail kite, and the Everglades.

Just as the snail kite was built to hunt the apple snail, Florida's only native apple snail was built to graze in the Everglades. The native snail's scientific name, *Pomacea paludosa,* invokes apples (*pom*) and marshes (*paludosus*). Our little marsh apples are the largest freshwater snail in North America. Their shells are commonly a deep brown, like polished mahogany.

I have only once gotten to watch a living snail motor around eating algae. It was in the Everglades National Park in 2009, at the opening of a culvert passing under a road. Its amorphous body was tan, and its tentacles waved around as it fed. Perhaps it was watching me; they have eyes on stalks. It's typically easier to spot egg clutches than the snails themselves. Back then, I would often see their white, pea-sized eggs, usually a double strand of twenty or so pearls, attached to vegetation above the water. And lots of empty shells.

For thousands of years, much of South Florida was marshy, and those white eggs on stems would have been a common sign of springtime. The snail kite's historical range covered much of the peninsula, even up into the Panhandle, but the Everglades was the heart of it (see figure 1.3).

The Everglades is itself an important and elusive character in the snail kite's story. Indeed, the Everglades National Park, which preserves a small part of the original Everglades, is undoubtedly our nation's most misunderstood park, even to people in Florida. Many people I speak to recoil slightly when I invite them to come along on one of my trips. The Swamp

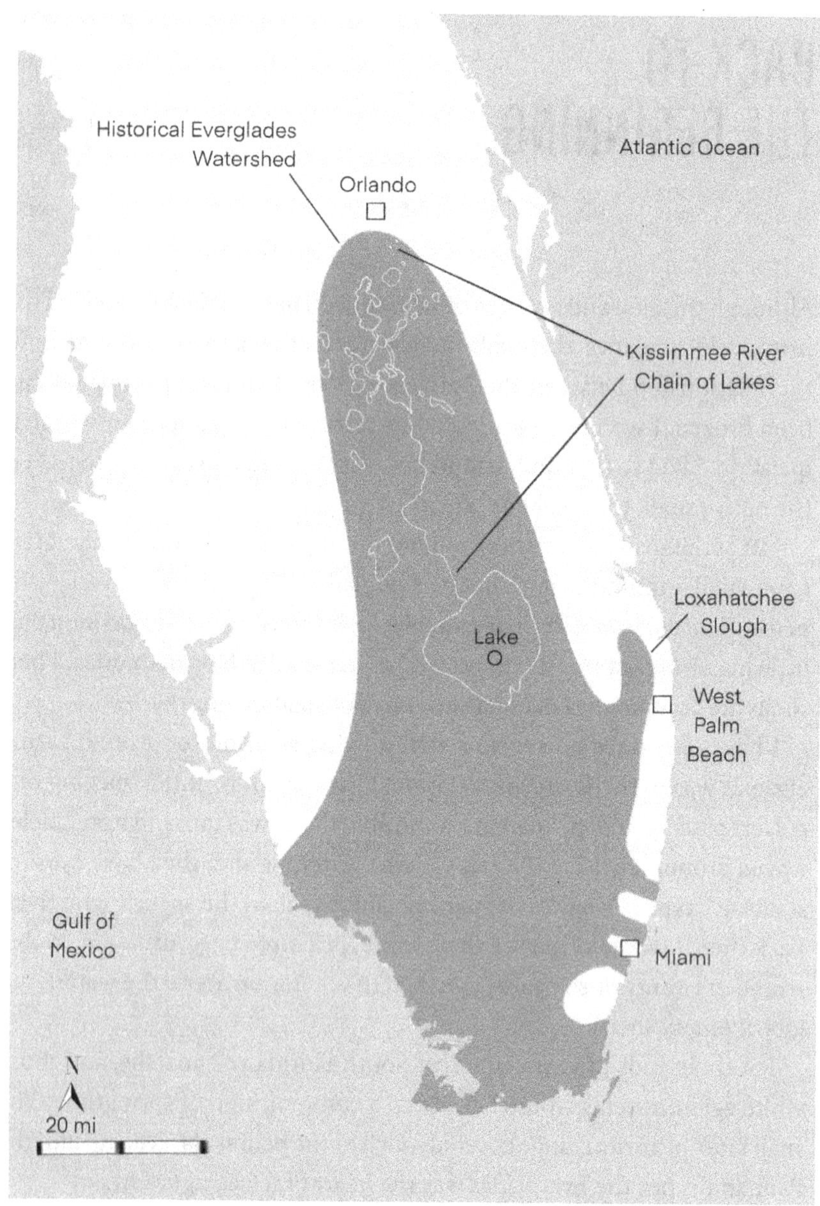


FIGURE 5.1. Historical Everglades watershed, as it was before about 1900, when the impacts of drainage and development escalated. The headwaters were just south of Orlando; some of the rain falling there would have flowed through the Kissimmee River Chain of Lakes, into Lake Okeechobee (labeled "Lake O") and south to the coast. Contemporary city locations are provided for geographic reference. Illustration by Hilary Flower.

Thing and the Creature from the Black Lagoon loom large in the popular imagination. And the prospect of being near alligators only adds to the Everglades' fearsome mystique. Many people who have been in swamps in other states may associate wetlands with bad smells and dark water. That is not the Everglades.

Before 1900, and for a few thousand years before that, if you were beamed into South Florida, you would feel the cool shock of water up to your knees or higher. Chances are you would be in a sunlit marsh dotted with lilies or other flowers. Lots of South Florida was marsh, studded with cypress domes and tree islands. If you looked up, you may well have seen a snail kite. The few written accounts we have of the Everglades before drainage simply describe them as common. It's not possible to know how many snail kites lived in the Everglades back then, but many thousands at least, possibly tens of thousands. A snail kite soaring above a marsh could easily spot apple snails in the clear, shallow water with sparse vegetation.

The unlikely conditions that turned the southern half of the peninsula into a shallow wetland is part of the snail kite's origin story. The Everglades is extremely young, from a geologic perspective. It only started to form about five thousand years ago. To put that into context, Indigenous people had been living on the peninsula for about eight thousand years before the Everglades even existed. When people had first lived on the peninsula, it would have been pinelands and grassy savannahs. But as the last ice age faded away, the climate became tropical: warm and wet. Five feet or so of rain fell most years, as it still does, enough to cover much of South Florida in two feet or more of water for much of the year. That's a lot of water.

It amazes me to think of people being here and watching, generation by generation, as the Everglades took shape. The headwaters were in what would later become Orlando. A drop of rain falling there would typically end up flowing about two hundred miles out to sea at the southern edges of the peninsula. Along the way it would pass through a maze of marshy lakes in the broad floodplain of the lazy, sinuous Kissimmee River, and then it would flow into Lake Okeechobee, which in turn overflowed to the south.

Running water will find slight low spots in elevation, and it will flow into them, cutting them deeper as time goes on, joining downstream into ever-larger channels. But the Everglades does none of that. The Everglades is the world's only large wetland dominated by "sheet-flow." There are other large and important wetlands, like the Pantanal in South America, but they are not sheet-flow wetlands. In the Everglades, the incredibly flat surface caused water to flow like a flat continuous sheet sixty-odd miles across and a hundred miles long until finally breaking into channels near the coast.

The Everglades' unique sheet flow arises from two flukes of its geology. First, it tilts very slightly southward. And the bedrock is flat from its long history as a shallow sea. The bedrock and the soil above it provided a vast surface flatter and more gradually sloping than what the finest engineers could have constructed: two inches drop of elevation per mile, for more than one hundred miles, from Lake Okeechobee to the southern coastline at Florida Bay. Extremely slight tilt meant that the water flowed very slowly, and there was not enough energy or relief to carve channels.

The Everglades' unusual sheet flow allowed it to break other rules. Vegetation cannot flourish where the current is too strong. The sheet flow was slow enough to allow plants to remain rooted. And the water was clear and shallow enough for sunlight to reach the bottom, allowing plants to photosynthesize below the surface.

In this way, a special confluence of geology and climate combined to form what Marjory Stoneman Douglas famously dubbed a "river of grass." In her book of that name, the opening words were: "There are no other Everglades in the world. They are, they have always been, one of the unique regions of the earth, remote, never wholly known."

The native apple snail evolved to thrive in the Everglades' precise pattern of seasonal water-level changes, low-nutrient water, and sparse vegetation. The Everglade snail kite is an important part of the college classes I teach because it offers a vivid example of how water affects wildlife. I'm not a birder or even a biologist; I'm an ecohydrologist. I study the interactions between wildlife and water in a landscape. Since the Everglade snail kite eats only one thing—the apple snail—and since the snail relies on specific water conditions, if you get the water right, the snail kite soars! Get the water wrong, and the kites vanish from the sky. The pivot point is the snails. The driver is the water. Simple.

For thousands of years, they coexisted: the snail, the kite, and the Everglades. Droughts came along periodically, causing the snail kites to search out wetlands that still had snails. But they had a lot of wetlands to choose from, so their population may not have had too many ups and downs.

And then everything changed. We can draw a line at the year 1900. That's about when drainage of the vast Everglades began in earnest. There were about 530,000 people in Florida in 1900; the population was about to explode. In the United States at that time, wetlands were considered wastelands, and draining them to benefit humans was seen as a moral and economic imperative. The Everglades was viewed with great animosity; the phrase "Drain the swamp!" is still in use today. The first half of the twentieth century saw rapid and unfettered development. People dug canals in the marshes and swamps, and the water drained out to the sea. Drainage created dry land for farms, ranches, roads, and towns in the growing young state. A large swath of marsh south of Lake Okeechobee became a hub of farming, and in the coming decades it became known as the Everglades Agricultural Area.

By the mid-twentieth century, there were five times as many people in Florida as there were in 1900. And about half of the historical Everglades wetland area was gone. The lowest point for the health of the remaining Everglades may have been in the 1940s. The Everglades was so dry that some years its soil caught fire in the dry season and smoldered through the wet season and into the next year. The Everglades National Park was created in 1947 to protect the southernmost part of the remaining wetland.

The loss of all that freshwater to the sea became a problem for the very farms and towns that had risen up on the newly drained land. Water supply became a pressing concern. In the 1940s, the US Army Corps of Engineers began constructing three Water Conservation Areas (WCAs) north of the Everglades National Park (WCA-1, -2, and -3; for location, see figure 1.2).

The rapid development across the state meant that by the 1960s, the main snail kite range in Florida had shrunk down to the southern end of the peninsula, with most of the surviving snail kites nesting in the WCAs and the western marshes of Lake Okeechobee.

The snail kite population may have dropped as low as forty, or even twenty-five, in the 1950–65 period. Although the numbers are uncertain, the possibility that the snail kite population ever approached double-digits hints that they could have slipped through our fingers into extinction

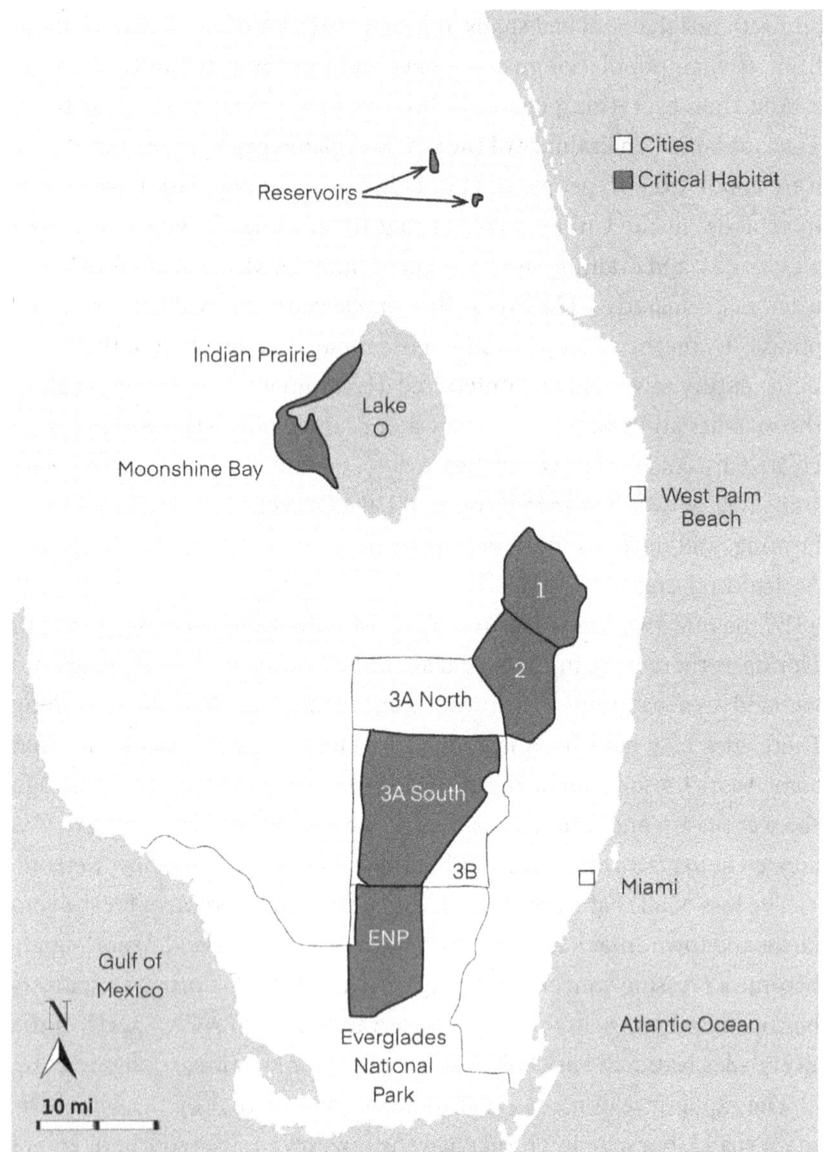


FIGURE 5.2. In 1977, the federal government designated Critical Habitat for snail kites to include their most important breeding areas in the WCAs, Moonshine Bay and Indian Prairie at the western edge of Lake Okeechobee (labeled "Lake O"), the Everglades National Park, and two reservoirs north of Lake Okeechobee. Under the Endangered Species Act, "Critical Habitat" was intended to designate the places essential for the species' survival and recovery. Illustration by Hilary Flower.

without anyone knowing it. It speaks to how vulnerable wildlife is to development that is not tempered by protections for wild areas. Habitat loss is a leading cause of loss of biodiversity, and it extends beyond the outright loss of wetlands to the degradation of those that do remain.

On the flip side, when the government started taking steps to protect wetlands, the snail kite populations responded. Under the name "Florida Everglade Kite," it was included on the first list of endangered species in 1967. The WCAs, and especially 3A, became the central hub for snail kite nesting, and the known snail kite population size increased by fits and starts in the coming years.

One of the ways in which the Endangered Species Act protects species is to designate "Critical Habitat" for them. Simply put, Critical Habitat for a given species is the habitat without which the species would go extinct. When Critical Habitat was registered for the snail kite in 1977, it was mainly the WCAs, the Everglades National Park, and the western marshes of Lake Okeechobee, areas where snail kites were most likely to nest and to be seen year-round.

A huge lake may sound like an odd place for a bird that needs shallow marshes. But Lake Okeechobee doesn't really look or act like a lake. Although it can be seen from space, very few Floridians or tourists have laid eyes on it. Driving by it, all you'll see is what looks like a long, grassy ridge, the 143-mile earthen dam known as the Herbert Hoover Dike, completed in the 1960s. You must drive up onto it to have a chance of seeing the lake.

And even then, it won't look like a lake. The eastern side of the lake is ocean-like: It has a rocky bank and gets deep rapidly. It looks like a seashore, complete with small clamshells, gently lapping waves, and open water to the horizon.

The western side of the lake doesn't look like a lake either. It either looks like a forest, due to the trees that grow along the edges of the rim canal, or a marsh. The western marsh takes many miles to make the gradual transition from the earthen dike to deep open water. A big section of that marsh is called Moonshine Bay, and another important marsh to the northwest is called Indian Prairie.

By 1979, the estimate of the snail kite population reached almost five hundred, still perilously low, but much better than forty. So much destruction had happened before we really understood what the Everglades was or how it worked.

The questions in 1979 were: Could Florida continue to develop without extinguishing its wildlife? Could we learn about snail kites and the Everglades fast enough to save them both?

And 1979 was the year when Dr. Steve Beissinger, the emeritus Berkeley researcher who had schooled me on snail kite adaptations, saw his first snail kite. At that time Moonshine Bay was sparse aquatic vegetation and open water from one horizon to the other, interrupted here and there by tufts of willow. Beissinger was just a rookie field tech doing a short-term job, but what he witnessed there pulled him into the world of snail kites. He stumbled on an adaptation that no one had realized was possible.

# THE CASE OF THE DISAPPEARING KITE

# 6

It all started early one morning in April 1979.

Beissinger was sitting on a piece of plywood fifteen feet up a tree in Moonshine Bay in western Lake Okeechobee. Next to him was Dr. Noel Snyder, then a biologist for US Fish and Wildlife Service (USFWS).

Their lookout offered a 360-degree view of sparse spikerush and fragrant water lilies. They were watching two nests about a hundred yards away. It was easy to see why snail kites had chosen to nest here. It offered just enough trees for nesting and perching. The snails crawled up the skinny green blades of spikerush in plain view through the shallow, clear water.

It is interesting to me that certain kinds of discoveries are made because of the arrival of a particular scientist with a particular approach. Snyder's predecessor had focused on finding all the wetland areas used by snail kites throughout the year. Where he went broad, Snyder went narrow and deep. The activity at individual nests was everything. Snyder wanted to determine whether snails were a limiting factor on snail kite population size, or if it was something else, like predators. To answer that, he wanted to see how hard it was for the snail kites to procure enough snails for themselves and their offspring.

So twice a week from dawn to dusk, Snyder and Beissinger watched two nesting pairs of snail kites, with a digital watch and binoculars in hand. They recorded how much time each bird spent foraging, perching, making aggressive chases, nesting, copulating, incubating, and feeding offspring.

That morning, Beissinger watched through his binoculars as the adult female kite flew off into the fog. I imagine the three young snail kites in her nest eagerly awaiting her return while Beissinger measured how long it took her. But the minutes stretched into an hour and beyond.

As the morning fog lifted, Beissinger timed the male snail kite as it continued feeding the young by himself. When the female had not returned by

FIGURE 6.1. The nest-viewing platform on the top of a tree in Moonshine Bay in the western marsh of Lake Okeechobee in 1979. Steve Beissinger (*left*) gazes out at the nesting kites 150–200 feet away. Another field tech, Gary Falxa (*right*) is recording information on kite behavior in his field notebook, the summer they worked as field techs for Dr. Noel Snyder of the US Fish and Wildlife Service. Photo by Noel F. R. Snyder.

the end of the day, Beissinger and Snyder were shocked. They concluded she must have died, picked off by a predator. That would be a little odd, because adult snail kites were seldom preyed upon except by the occasional great horned owl.

But deliberate abandonment would be even odder. For a species to endure, its members must collectively produce offspring that will survive long enough to reproduce themselves. For these two kite parents, that mission translated to weeks of foraging snails for their babies. With three babies, that meant forty-five snails a day, plus each parent needed fifteen for themselves. To go it alone, the mate left behind with three nestlings would need to secure sixty snails a day, about one every twelve minutes. If not, the nest would fail.

Beissinger became curious as to whether the mate left behind had clocked his predicament. The stopwatch showed that he was delivering snails to his nest more frequently: He was aware of his situation, and he had decided to rear the young alone.

A few days later, it happened again. One of the parents at the other nest that Beissinger and Snyder were observing flew away and never returned. This time it was the male who disappeared, leaving the female behind.

Usually a discovery is made when something new shows up. This discovery was made when two birds went away. There was no case in the scientific record of snail kites deserting their mates. That alone was surprising enough. But could it really be that either the male or the female could desert?

Two-parent care is mandatory among bird species, with fewer than 9 percent engaging in desertion. Eggs and young birds need constant incubation to maintain temperature, not to mention protection from predators. Typically, the mates take turns guarding the nest and foraging for food. For the few bird species known to desert, the deserter was always a consistent sex. Young magnificent frigatebirds see the last of their father after they hit three weeks of age. Very rarely, a species is known for females deserting. The female boreal owl will sometimes desert if food supply is high. There is a shorebird in Russia called the red-necked phalarope; the female sometimes abandons her mate at the egg-incubation stage. Female European starlings sometimes desert, but not before their nestlings are nearly ready to fledge.

Beissinger investigated as many snail kite nests as possible. To his surprise, this behavior proved the rule rather than the exception. Of the thirty-six nests they studied, twenty-eight were deserted by one of the parents, and it was about equal between male or female.

If a predator were to blame, the timing would be random: Solo parenting would have been just as likely with eggs or tiny hatchlings in the nest. But Beissinger found that the young in solo-parented nests were always around four to six weeks old, about halfway between hatching and fledging. This is the age threshold when baby snail kites no longer need a parent to keep them warm; they could maintain their own body temperature.

"So that's when we realized," Beissinger told me, "that these probably were not instances of mortality." He raised his eyebrows with his hands palm-up in remembered bewilderment: "The kites had a flexible system of parental care. It was shocking. And we wanted to know: How often does this happen? Is it known in other raptors? Noel is a raptor guy, and even he was shaking his head about this, trying to figure out why? What's going on here?"

Perhaps the most astonishing part of all: It was a successful strategy—even for nests with three abandoned young! The only failure of a solo-parented nest that summer happened late in the nesting season. Solo parents worked extremely hard, bringing their young only half a snail less per hour than paired mates, and they expended twice the energy of nonbreeding kites.

Beissinger and Snyder were the first to publish on what they dubbed "ambisexual mate desertion" in birds. Even now, decades later, fewer than 1 percent of bird species are known to engage in this freewheeling reproductive strategy.

This discovery changed the course of Beissinger's life. He started a PhD program to find out everything he could about the snail kites' ambisexual mate desertion. Most of all, he wanted to get into their heads and figure out how a given snail kite made the decision whether to desert or not.

One thing became clear early on: Overlapping nesting gave the snail kites a big boost in reproductive output. At the end of 1980, the snail kite population estimate was up to 651, 150 more than the year before.

When Beissinger finally launched his own graduate fieldwork, in the spring of 1981, snail kites were hard to find, and none were nesting. He was desperate to find nesting snail kites. When he did finally find a few, they defied his expectations yet again: No kites deserted their nests!

Sorting through his data, Beissinger figured out that desertion was an ingenious adaptation to apple snails. Snail populations went up and down with cycles of high rainfall and drought. Snail kites had to make the most of good years (desert) and get through bad years (don't desert). Crunching the numbers, Beissinger found that he could predict the occurrence of mate desertion based on how far the kites had to fly for snails. In the years when snails were scarce and kites had to fly up to 3.5 miles between meals, desertion was rare. The switch flipped when the kites flew less than 1.5 miles to forage for snails; nearly all nests were deserted by one of the parents. On those "good snail years," they overlapped their nesting efforts and extended the nesting season as long as the snails held out.

The peak breeding season for kites was February through July. But Beissinger found that if the water conditions allowed for it, the snail kites could keep at it for up to ten months. He said, "The long breeding season and mate desertion gives a deserter a way to increase the number of young that it raises in a nesting season."

The strategy proved to be calculated, multifaceted, and subtle. A future deserter would often slow down their snail deliveries, as if to see whether their partner would pick up the slack. If more single males were in the neighborhood, females were more likely to desert and vice versa. Nests with fewer young were more likely to be deserted and were deserted sooner.

Whenever Beissinger could, he would band the young kites. This involved carefully holding a nestling and attaching a PVC band with a combination of colors, letters, and numbers that would allow the individual to be identified in the future. They also attached a metal band with a unique number that the USFWS issued to bird banders.

Noel Snyder had adapted their airboat to include a wooden "work desk" for banding birds, setting up a spotting scope, and taking notes. They towed the airboat behind the field truck, and the banding desk raised some eyebrows. Beissinger said, "It was quite unusual. Locals around Lake Okeechobee thought it was a 'pulpit' and wondered if we were 'preaching' to the animals."

FIGURE 6.2. Steve Beissinger is at the improvised work desk on the airboat, rotating a pole with an antenna attached to track radio-telemetered snail kites in 1981. Photo by Noel F. R. Snyder.

While handling a bird for banding, Beissinger took measurements on them and sometimes attached little backpacks with solar-powered radio transmitters, an earlier version of GPS trackers now used by ARCI. To pick up the signal, they had to hold an antenna and cruise around on an airboat, or fly over in a small aircraft with the antenna attached to the struts. Radio telemetry and re-spotting banded birds in new places allowed Beissinger to

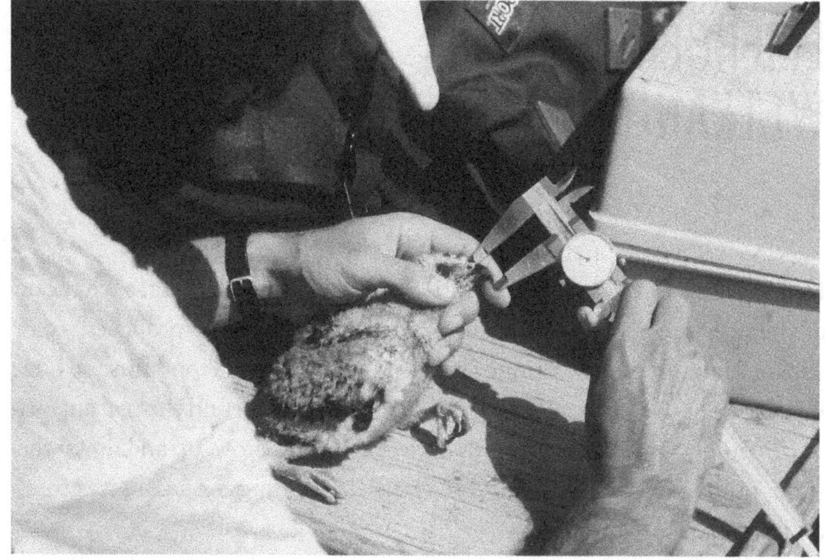

FIGURE 6.3. Steve Beissinger uses a caliper to measure the bill of a nestling snail kite in 1983, on the airboat's work desk. Photo by Noel F. R. Snyder.

confirm that the deserters went on to start a new nest with a new partner. Beissinger said, "Successive nesting attempts may be as near as one hundred yards or as far as one hundred miles." He followed one telemetered female who, he said, "was incubating eggs two weeks before her former mate was finished caring for the original young!"

When I had first started digging into snail kite adaptations, I had thought of it in terms of physical adaptations: bills, talons, and wings. Beissinger was showing me that their behavioral adaptations were at least as important, if not more so. For millennia, mate desertion allowed this specialist predator to make the most of a prey item that can have ups and downs. The snail kites' knack for surprising researchers was a measure of their rapid adaptability.

And it turns out that the disappearance of kites when Beissinger was trying to start his graduate research in 1981, which had seemed like such a setback, was a clue to another behavioral adaptation. Another discovery by absence. And this one goes a long way to explaining snail kites' having a nesting supersite up at Paynes Prairie in recent years.

# EMERGENCY WETLANDS 7

Had the snail kites disappeared into a black hole in 1981? That's what one of the researchers trying to conduct a population survey proposed as an explanation for his low count of 109 kites that year. And that must be how it felt to Steve Beissinger that spring. But he needed to find the real answer so he could start fieldwork for his dissertation. He drove up and down the state, pulling an airboat behind his truck, searching for snail kites in as many wetlands as he could.

Meanwhile, snail kites were showing up in unusual numbers at the Loxahatchee National Wildlife Refuge, in Palm Beach County, just southeast of Lake Okeechobee (shown as WCA-1, figure 1.2). Jean Takekawa was a newly minted wildlife biologist in a field office for the US Fish and Wildlife Service (USFWS). The sudden influx of this endangered species got her attention.

And then within weeks, the snail kites vanished from there, too. Where had they gone? And why? As the spring wore on, the answer to "why" started to become clear. By May, there was no longer standing water in the Refuge. By July, the top layer of muck was cracked and dry.

Takekawa's concern for the snail kites deepened further when Beissinger swung by the Refuge to trade information about the missing snail kites. He told her that when the biggest snail kite stronghold, Water Conservation Area (WCA) 3A, became too dry for an airboat, he had looked by helicopter. By June there were no kites. Lake Okeechobee was doing a little better for a while. In June, he had found two snail kite nests in Moonshine Bay. Each fledged one young. Eventually, the stragglers at Lake Okeechobee were clustered near a canal. Some resorted to eating turtles. The drought had driven the snail kites from their primary habitat. But where had they gone?

Takekawa told Beissinger that she was getting calls from across the state reporting sightings of snail kites. Call by call, the public were providing her

FIGURE 7.1. This sidebar appealing to the public to call in to a snail kite hotline with kite sightings appeared alongside an article about the plight of the snail kites titled "Battle for Survival, Rare Bird Imperiled by Drought and Man" (*Fort Lauderdale Sun-Sentinel*/TCA, August 17, 1981). Used with permission.

## *You can help keep Florida's kites airborne*

**You can help the endangered Everglade kite by reporting all sightings to the U.S. Fish and Wildlife Service staff at Loxahatchee National Wildlife Refuge at (305) 732-3684 in Boynton Beach.**

**The kite is a relative of the hawk but different in many ways.**

**Its beach is longer, more curved and yellowish or reddish at the base. The beak is black at the tip.**

**Males are dark slate gray, although their heads and upper backs may be a lighter gray. Females are a mixture of mottled brown and gray on white or cream. A white line runs over the female's eyes. Both the male and female have white patches at the base of their tails.**

**Kites nest and feed in marshy areas. Their nests generally are built close to the water in sawgrass or similar vegetation.**

with surprising and far-flung indications of where the snail kites were.

Takekawa decided to leverage the public's observations to solve the puzzle, and the Kite Sighting Hotline was born. She put out advertisements and press releases to over two hundred media outlets around the state, instructing the public on how to recognize snail kites and entreating them to call into the Refuge with their reports. Takekawa was ahead of her time. It would be several years before the term "citizen science" was even coined. The field of "conservation biology" did not yet exist.

Over the fifteen months of the drought, she fielded 155 calls from biologists, game wardens, birders, naturalists, hunters, and farmers. She asked each caller a series of questions for verification, and she went in person to see for herself when she could. Most sightings proved credible.

I was excited when I got to sit down with Takekawa via video at her home in the Pacific Northwest. She had shoulder-length silver hair, and she sat at a desk with a big bay window behind her. As we talked, her expression moved fluidly between a furrowed brow, and eyebrows raised with joy and laughter.

"The kite sighting program turned out to be pretty darn powerful," she told me. "We were fortunate that snail kites are so distinctive."

The stakes were high: Like so many droughts before and after, half of the snail kites would die. One woman who tried in vain to save an emaciated snail kite told a newspaper reporter, "You feel like sitting down and crying your heart out because they are so beautiful and there are so few left. It's just a beautiful bird."

Still, with each call, Takekawa got another datapoint of kite survival, and she marveled at the geographical spread the hotline revealed to her. During the height of the drought, desperate kites flew hundreds of miles in search of snails. Snail kites were sighted northwest in the Panhandle and northeast in the Jacksonville area. For some counties, it was the first ever record of a snail kite. The kites were returning to parts of the state they had probably occupied in the pre-1900 era, before the state had become drained and developed.

The snail kites continually surprised Takekawa about the area she thought she knew around the Refuge. In the 1980s the City of West Palm Beach's wetlands were being rapidly converted to urban sprawl. One day, someone called to report snail kites at a new well-to-do housing development near the Refuge. Takekawa looked at me with wide eyes, saying, "We considered that place lost. We had given up on that area." But when she checked it out, she was thrilled to see snail kites foraging in the retention ponds there. She was curious how they were getting enough snails to eat, given the modest size of the water bodies. Once, she counted thirty-five kites foraging at Wellington Pond, a nearby seventy-four-acre water treatment impoundment. This was the highest density of snail kites ever recorded foraging over a sustained period.

It turned out Takekawa's local area in West Palm Beach saved the snail kites that year. She was intrigued that most of the wetlands they relied on were too small for snail kites to use on a normal year. She realized that this swath of wetlands, which she dubbed the "east coast corridor," was part of the snail kites' birthright. These wetlands were remnants of a large wetland called Loxahatchee Slough, a northeastern arm of the Everglades watershed in the pre-1900 pre-drainage era (see figure 5.1). She suspected that the snail kites had always turned to the Loxahatchee Slough during drought years, and she wondered if there was some way in which this knowledge was passed down. The east coast corridor made the perfect drought-refuge for many reasons. First, it gets the highest rainfall in South Florida, then and now. And second, it is unusually good at holding onto

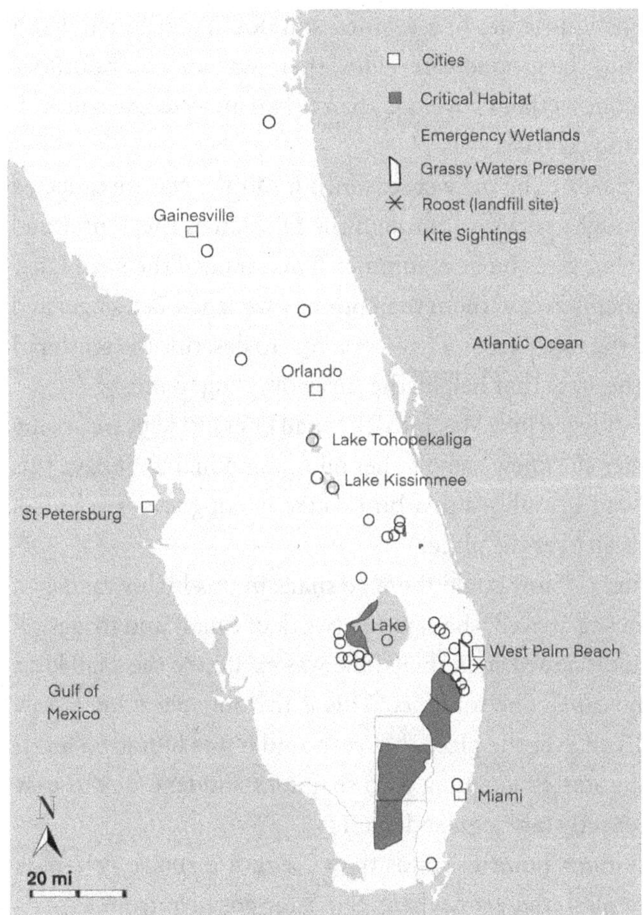


FIGURE 7.2. Emergency wetlands where snail kites survived during two major droughts in the 1980s (1981–82 and 1985). Many sightings were reported from the hotline. The concentration of sightings near West Palm Beach corresponds to what Jean Takekawa dubbed the "east coast corridor" (which in turn corresponds to an arm of the historical Everglades watershed known as Loxahatchee Slough; see figure 5.1). This region included a water catchment area that later became Grassy Waters Preserve, and a large roost that Jean dubbed "the landfill site." Snail kites also heavily used Lake Toho and Lake Kissimmee. Note the lack of overlap between these emergency wetlands and the snail kites' Critical Habitat designated in 1977 (see figure 5.2). Illustration adapted by Hilary Flower from Takekawa and Beissinger, "Cyclic Drought, Dispersal, and the Conservation of the Snail Kite in Florida" (1989).

water. Like Paynes Prairie, its fine-grained soil doesn't let much water percolate down into the groundwater. Plus, there was no major outflow. And it still had more wetland coverage than urban areas to the south. It was the perfect drought oasis.

Summer is supposed to be the wet season, but Mother Nature canceled it in 1981. The drought persisted through the fall, across the winter and spring dry season, on into the next summer. This strained the snail kites' ability to sustain themselves without their primary wetlands. Beissinger and Takekawa coined the term "emergency wetlands" to describe the scattered wetlands across the state that helped the survivors cling to life.

As the drought dragged into May 1982, the snail kites just kept surprising Takekawa. A farmer she knew came to her office announcing, "I have this bird in my field, and I think it's that bird you're looking for." He added, "And there's snails all over the place!"

Takekawa thought, "How could there be snails in this highly managed farm field?" When she arrived, she saw a network of canals and irrigation ditches lined with discarded snail shells. She was excited to see snail kites foraging along the canal. She told me, "This is the only place I ever saw piles of snail shells all over the place that year. And it was literally piles! It was shocking." She started laughing with shaking shoulders. "It was easy to detect. Didn't exactly take a great detective!"

The shells had much pointier spires than the native snails. Pale-pink egg clutches were plastered everywhere. She later got help from experts to identify this snail as *Pomacea bridgesii*. She looked hard but couldn't find native snails or their eggs on the farmer's property. It turns out that Takekawa witnessed the first and only recorded *P. bridgesii* population boom in Florida. It foreshadowed the current situation, where snail kites are relying on a different non-native apple snail. The fact that *P. bridgesii* was the same size as the native snail undoubtedly helped the snail kites switch to them without skipping a beat.

Around the same time, Beissinger struck gold. He found snail kites in two large lakes about sixty miles north of Lake Okeechobee in Central Florida, in the Kissimmee River Chain of Lakes Area. West Lake Tohopekaliga (Lake Toho, for short) and Lake Kissimmee were deep enough and large enough that they still held water despite the drought. Water meant snails, and snails meant kites. Together these lakes hosted a quarter of the known kite population in 1982. Beissinger was shocked and delighted to find a few nests at these two lakes, the first ever recorded there, and the only nests that year. Lake Toho produced nineteen fledglings, and Lake Kissimmee produced five. To have survived the year was an accomplishment, to have reproduced was over the top.

In June 1982, the drought ended all at once when a huge tropical depression sat over South Florida. "It rained like hell for multiple days," Beissinger recalled. Lake Okeechobee's western marsh and the WCAs rapidly reflooded. And the kites returned home almost immediately.

"It's amazing how quickly the kites moved," Takekawa said. "They disappeared from these marginal places. It's mysterious, but"—she nodded with a big smile and a sparkle in her eyes—"I guess they knew." She pointed out that if they went back prematurely, they could starve to death. When snails were scarce, the kites had to limit how far they flew, to conserve energy. She said, "They would have to assess, 'Is it worth the flight to go back to the park or their usual wetland?' That could be a fatal flight if you don't get snails." She went on, with wide eyes, "But you know, you see one show up, and all of a sudden there's seven. What's up with that? The social nature of kites, it's so mysterious." She laughed with affection and admiration.

This reminded me of how Dr. Caroline Poli had been surprised at Paynes Prairie in 2016 to see first one and then several snail kites.

I pressed Takekawa, "Do you have any guesses at all about how they are notifying each other?"

She shrugged with a big smile. "There must be more communication going on than we know. They're somewhat social, and they sometimes roost in large groups. So . . . ?" She shook her head.

Takekawa and Beissinger had revealed the snail kites to be nomads: Their habitat was wherever the snails were, even hundreds of miles away from their typical haunts. This was an important finding, with big implications for their conservation. When Takekawa and Beissinger wrote up their findings, they made a plea for a policy change. They realized there was a fundamental problem with the Critical Habitat established for snail kites in 1977. For some animals, it's pretty easy to identify the areas that stand between them and extinction, because they are always there using them. Similarly, it's fairly straightforward to designate Critical Habitat for migratory birds, because they tend to have two or more main sites they move to seasonally, in a fairly predictable manner.

Nomads require a different approach. Beissinger and Takekawa asserted that to prevent extinction, three tiers of sites needed protection: the snail kites' primary nesting grounds (e.g., 3A and Moonshine Bay in Lake Okeechobee), their secondary wetlands (where they go next if the primary wetlands dry up), and then the emergency wetlands, rarely used by kites except in drought years. The pair foresaw snail kite extinction without their emergency wetlands, and the feverish pace of development was a big threat. The problem was that when Critical Habitat was registered for the snail kite in 1977, only the kites' primary nesting habitats and some secondary habitats were included.

In a paper the duo published in 1989, Beissinger wrote a section arguing that the snail kites' Critical Habitat needed to include at least the east coast corridor, Lake Toho, and Lake Kissimmee. He pointed out that droughts were a normal part of the Florida climate, coming and going on a five- to seven-year cycle. He argued that, although snail kites had weathered droughts in the predevelopment past, that was when the Everglades was twice as big, and wetlands were abundant through much of the state.

Further, Beissinger noted that by the 1980s, droughts were becoming much more of a threat to the snail kites for several reasons. Widespread drainage had lowered the water table, which caused wetlands to dry down sooner. Increasing water demands made water that much scarcer in wetlands during droughts. More troubling still, droughts were becoming more severe and frequent. I was surprised that the effect of climate change had already become noticeable back in the 1980s. I started to worry about how climate change would affect them in the coming years, but I set that aside for later. Beissinger warned that the snail kites' emergency wetlands would only become more crucial for species survival in the decades to come. He noted that kites returned to specific habitats in subsequent droughts, like Lake Toho, Lake Kissimmee, and the east coast corridor. These places needed protection.

Beissinger also made a key observation, which may have implications for the recent decline in native apple snails. Two drought years in a row, like they observed in 1981–82, can wipe out an entire snail population. This stems from the fact that it takes the native snails about a year to mature enough to lay eggs, and they die soon after. They can survive dry conditions for several weeks, but an extended drought can kill off a great proportion of adult snails before they are able to lay eggs. If the following year is again poor for snail reproduction, there may be few survivors. It can take up to four years of rainy weather for the population to rebound. I wondered if this "lag effect" could help explain the scarcity of apple snails in the Everglades in the last couple of decades.

Beissinger wrote: "This scenario—cyclic drought, a falling water table, and fewer drought-related habitats in the east coast corridor—paints a bleak prognosis for the snail kite in Florida. If the snail kite is to survive, we believe that as many drought-related habitats as possible must be preserved and protected from alterations in hydrology." It was one thing for Beissinger to write that, as an academic, but Takekawa worked for a government agency. When we talked, she was surprised that it had gotten past her supervisor. But later she dug into old files and found a handwritten note from her supervisor, giving his blessings. He added, "I see nothing controversial here."

I listened, riveted, to Takekawa talk about all of this. I told her that snail kite Critical Habitat had still not been updated. She said that many of the emergency wetlands that she found in 1981–82 had undoubtedly been lost.

She and I talked about how her discovery of the emergency wetlands and the kites' nomadism went a long way to explaining the snail kites' recent big shift north. After all, snails have been scarce in the Everglades for more than a decade now, although it's not due to weather this time. Snail kites have gone searching for snails just like they did in 1981, and it has landed them in unusual places, including Paynes Prairie hundreds of miles north. The native snail emergency in the Everglades is stretching into decades. It is not clear this time when or if they will ever be able to return to 3A and their other primary wetlands.

Then Takekawa's face lit up. She wanted to tell me about "an even bigger snail kite discovery."

# MIRACLE ROOST

# 8

After rains extinguished the drought of 1981–82, rain kept coming for the next two years. Great for snails, great for snail kites. Steve Beissinger watched them have multiple broods, deserting their mates to start new nests. By the end of 1984, the snail kite population estimate had rebounded to the predrought numbers, about 650.

Jean Takekawa's big discovery happened during the next drought. She said, "The 1985 drought, that was the biggest deal. The kites disappeared again."

This time her Kite Sighting Hotline was a well-oiled machine. In March, she ramped up the publicity to gain new volunteers, and more than half of the last crew helped again. As Takekawa vetted the 299 calls, she saw that not only did the snail kites reuse many emergency wetlands from before, but they revealed new ones.

There was one that she would never forget.

It started with a call from some birders seeing a few kites in the West Palm Beach area, northeast of the Refuge and about forty miles due east of Lake Okeechobee. She was not even aware of any wetlands in that particular spot, and none were visible from outside. The caller said there was a wetland there that was closed to the public. Takekawa met them at a large pond on a dirt road, and they could see some birds at a distance: a few kites, as well as some wading birds. It was in the east coast corridor, but this one was new to her.

Takekawa started returning there in the evenings for roost counts. When she learned that a landfill was planned less than a thousand feet from the roost, she started referring to it as the "landfill site" (for location, see figure 1.2). The ponds were abandoned shell pits where the fossil shells and limestone bedrock had been excavated for road construction and aggregate for concrete. In Florida, the landscape is so flat and the water table is so high that any break in the bedrock rapidly fills with water. The mining left a corrugated pattern of ten-feet-deep elongate pits separated by piles of discarded overburden. Willow and an invasive plant known as Brazilian pepper grew thick and tall on these narrow islands.

One June evening out there in 1985, Takekawa spotted some kites flying at a distance. Through her binoculars she watched them fly over to one of the islands at the far side of the pond.

Then five more.

Then ten more. Twenty. Takekawa told me with wide eyes, "The snail kites just kept coming in, and coming in. I started counting, and it was just ridiculous. There were wading birds coming in, too: herons, egrets, spoonbills. I could see there were nesting colonies of wading birds there. And the kites just kept coming in. There were too many for me to count, especially at that distance. I realized that there was some crazy, ridiculous number of kites using that area, safely more than a hundred."

"It was a revelation," she said. "Snail kites had seemingly disappeared in response to the drought, yet here they were hiding in plain sight. No one had guessed or predicted this response. The kites had found their own solution to help them survive the drought."

She stopped trying to count them. She just watched in awe. She couldn't wait to share her discovery. But for that evening, she savored the moment of knowing something that no one else knew, seeing something that no one else had ever seen.

The next day, she mobilized. She recruited another US Fish and Wildlife Service (USFWS) biologist, Ellie Van Os, who was thrilled to get away from her desk. Takekawa wanted to get closer to the roost, but gas-powered boats were barred due to water-quality concerns. She got special permission to bring in a canoe, and they paddled in. Many thousands of wading birds were nesting in the islands. Takekawa and Van Os saw several kites flying around, and when they spotted their hub, they paddled to it.

She brightly said, "And then, most exciting of all, we saw a nest!"

Takekawa assumed that snail kites couldn't be nesting that year; they were focused on survival. But then she spotted a dense concentration of sticks in thick Brazilian pepper. She watched the male snail kite parent tend to the nest. And then she saw a brown bird fluttering at the nest—a fledgling!

FIGURE 8.1. Jean Takekawa in a canoe at one of the forested islands in an abandoned mining pit in West Palm Beach. She had come that evening in 1985 to count the kites as they came in to roost as part of her work for the US Fish and Wildlife Service. She is gazing up at an adult male snail kite (arrow points to him atop a high branch) that was tending a fledgling near a nest (out of view). Photo by Ellie Van Os.

Takekawa said, "The discovery that a nest would have been attempted—and was even successful at fledging a juvenile!—seemed so unlikely in those extreme drought conditions. It showed a resilience and adaptability by snail kites that gave me more hope."

Through her binoculars she saw that the male had a band on its leg. It was one of Steve Beissinger's bands, from 1979. Takekawa marveled, "The excitement of finding a nest first of all, of even seeing it. But to know which individual that was, I mean, it was just an amazing discovery." She explained, "Resighting a banded bird in the wild like a kite is a rare piece of information, especially in drought conditions. It helped us understand that this was a relatively old, veteran bird, but also that it was nesting quite far from the original banding site. For me, it showed the ability of this bird to survive and adapt, despite very stressful conditions."

Takekawa went back three days later with another biologist and counted 226 kites. She could tell that there were even more kites that they couldn't count without more help to cover all directions. Four days later, she came back with three additional observers. They broke into two teams of two each, and split up counting by different directions, using landmarks like certain islands and distinctive trees. That way, they could be sure not to double count. The teams were close enough to each other that they could still be in communication. For each team, one person called out as kite came in to roost, while their partner wrote the number down (e.g., 1–3–2–4, etc.).

I asked Takekawa if it was difficult to count the kites with them flying around. She said, "Fortunately, once they landed and roosted, they usually stayed put, instead of just flying around, flying around. That would have been impossible."

When she started to compile their numbers at the end, Takekawa was prepared for a high number, but not as high as what they counted. She said, "We counted 372 snail kites!"

I was stunned.

"We could hardly believe there were that many snail kites roosting in one place, in one relatively small wetland, one that we had never really known about," Takekawa said. "At that time, it was the largest roost ever found. I have to say that was the most earth-shattering thing of that whole hotline program for me."

At the end of 1985, the known snail kite population was just four hundred. Takekawa's landfill roost hosted most of the drought's survivors. Talk about an emergency wetland!

And it called her attention to another hidden wetland, one just as important and crucial. While the abandoned mining pits provided safe roosting, the ponds were too steep-sided and deep to support emergent vegetation and snails. Takekawa recalled, "There was not a lot of evidence of snails. We didn't see much in the way of snail egg clusters. We knew there was no way that 372 snail kites were feeding just in those mining pits." Takekawa deduced they were foraging in a nearby wetland. But where?

They started to take careful note of where the kites were coming from. Again and again, the kites were coming from the west. The western boundary of the landfill site was a water catchment area: a broad, shallow wetland that provided drinking water for the City of West Palm Beach. It was also closed to the public, so Takekawa had been unaware of it. "It was a very protected spot," Takekawa told me. "I thought that was pretty smart of the kites." By day, they foraged in the water catchment area; by night, they roosted in the landfill site.

Finding this big roost was bittersweet for Takekawa, because she would be transferring to another refuge soon on the other side of the country. Her work at the Loxahatchee National Wildlife Refuge had confirmed that she wanted to devote her life to saving endangered species and wild areas. She felt that the 372-kite roost was a great going-away gift from the kites. And she wanted to give a gift back to them if she could.

She called for the City of West Palm Beach to designate their water catchment area a kite sanctuary. And she tried to prevent landfill construction from destroying the 372-kite roost site. She reached out to the Solid Waste Authority (SWA) of Palm Beach County, and the planners invited her to tour the site with them. She pointed the roost out, explained how important it was to the kites, and urged them to protect it. She made sure her supervisor at the USFWS knew its importance as well; it was all she could do. And then she left Florida.

Since 1985, the City of West Palm Beach has doubled in population, and so has the state. I asked Takekawa if she knew what had become of the water catchment and the landfill roost. She didn't know what happened to the landfill roost, since the landfill construction happened after she left. I could see that there was indeed a landfill on that property now. She decided to look up the water catchment area before our interview, to see if it was still there. She said, with a pained expression, "I was just dreading looking it up because I was thinking, 'It's gone. It's developed. It's housing.' I looked it up, and I saw that it's called a preserve now. It's called Grassy Waters Preserve, set aside by the City of West Palm Beach. They have trails and a boardwalk. In the picture I saw online, it looked beautiful. And the little write-up specifically mentioned that it's important for snail kites." It seemed like the next best thing to being named a snail kite sanctuary, as she had requested so long ago.

She beamed, saying, "I am amazed that it is actually still there all these years later, and it is being nurtured. They have a visitor center there and they have education programs."

Feeling a sense of hope bloom in my chest, I blurted out the question that percolated below all my questions about snail kites. "Does this kind of thing give you hope? Do you have hope for the future of imperiled animals?"

Takekawa didn't skip a beat. She furrowed her brow, saying, "I have to. I have to. I mean, a lot has been lost, but there's still a lot left to protect. There's still a lot of difference that people can make if they work together. I truly, truly do believe that there's some really powerful things people can still do."

She would know. After leaving Florida, she had transferred first to the San Francisco Bay National Wildlife Refuge. When she arrived, a non-native red fox was sending the California clapper rail to the brink of extinction. Under her leadership the rails rebounded. In Nisqually National Wildlife Refuge in Washington State, Jean Takekawa headed the largest tidal marsh restoration project in the Pacific Northwest.

Takekawa said, "When I look back, I realize I was so lucky to grow up as a biologist in an atmosphere that was kind of freewheeling. In a way, it was like, 'All bets are off! Go for it!' And I worry about young people today. It has to look bleak. I hope young people will still go into the field of conservation and be inspired to know that they can prevail. I want to tell them: Look for the compromises and the ways to achieve your goals. But don't be discouraged. Persist. Persistence is everything. My mantra is: There's always a solution. It may not be perfect. Don't be discouraged by what seems like overwhelming odds."

She laughed. "When I would be frustrated, the seasoned hands would tell me that politics and administrations come and go, and that the agency and the efforts to protect persist. If young people go into conservation, they get to live their life following a passion. That's huge. Is there anything better than to do something you believe in?"

Before meeting Takekawa, I had not realized how much a single wildlife biologist could accomplish, and in just the few years she was in Florida, much less California. We might not have snail kites in Florida today if she had not found the roost, and other important hidden refuges in the urban and agricultural landscape. She and Steve Beissinger showed that protecting wetlands in the human-dominated landscape was just as important for snail kite survival as protecting the Everglades.

After our interview was over, I looked up Grassy Waters Preserve. When

I saw that they had guided kayak tours, I knew I had to go. I didn't know yet what piece it would contribute to the snail kite puzzle, but I knew it was the next step the snail kites were leading me to.

My first conversation with folks at Grassy Waters upended my assumptions about its relationship with kite history.

# THE SECRET OF GRASSY WATERS

# 9

Vera de Chalambert answered my call to Grassy Waters Preserve. When I mentioned how I heard about Grassy Waters, her tone shifted. She said she could hardly breathe. She put me on speaker phone and introduced me to Lauren Butcher, one of the preserve's environmental education coordinators. I heard Butcher say, "Hello!" from the background.

Here's the thing: No one there knew that Grassy Waters Preserve had helped save the snail kites in the drought of 1985. Nor had they heard of a 372-kite roost. After Takekawa had left the Refuge, a decade passed before the nature center was built and its education program began, so the lore was not passed down. The preserve's emblem was a snail kite, but not because it was recognized as a snail kite sanctuary. Indeed, their original symbol was a wading bird. When they had to change it a few years ago, they chose the snail kite simply to reinforce the fact that the preserve was part of the Everglades.

De Chalambert had written a children's book about a snail kite at the preserve, but again it was not because of the special history there. As de Chalambert had sat wondering what animal to focus on for her book, a snail kite flew past her window. That seemed like her answer. The book is *Sam the Snail Kite and the Secret of the Everglades*. A young kite sets off to ask wetland dwellers such as Papa Cypress, Mother Sawgrass, Uncle Gator and Aunt Otter for the secret of keeping the Everglades alive. I found myself eager to learn what the answer would be; I was on a similar quest, after all. Finally, Sam asks an Indigenous person, who responds, "Oh Sam, I DO know the secret! The Everglades has been home to Indigenous people since time began. Our elders know that WATER is the heart of the Everglades. So, the secret of keeping the Everglades alive is PROTECTING THE WATER! We must all become water protectors." The book gives numerous examples of what that looks like, such as conserving water by fixing leaks and not leaving the water running. "Grandma Cloud" suggested, "Protect natural areas!"

The following Sunday, I drove from St. Petersburg, Florida, to the other side of the peninsula with one of my students, Kait Kennedy. Lauren Butcher welcomed a dozen of us to a covered deck by a lily-covered wetland. She gave us a safety talk, a life vest, and a paddle, and soon we were lowering ourselves into kayaks. We paddled behind her, away from the deck and civilization.

It felt good to sit low in the kayak, at the wetland's surface, like another wetland creature. I slipped my paddle into the clear water between fragrant water lilies with fat green disks for leaves. We paddled through patches of sparse spikerush and patches of spatterdock with its yellow bulbous flowers reaching several inches into the air. We paddled single-file past a narrower section where a large gator watched placidly as we passed by.

Up ahead, near a cypress dome, Butcher turned her kayak to face the assembled group. She said, "Look around you. This is what the Everglades looked like before the development of the last century. Grassy Waters has never been developed. It has been protected all this time." She beamed. "It's like a window back in time."

I thought about how Jean Takekawa had realized that this "water catchment area" and the other wetlands in the east coast corridor were fragments of the Loxahatchee Slough (see figure 5.1), an arm of the original Everglades watershed. We listened as Butcher explained that in the 1890s a developer had purchased the wetland to supply water to West Palm Beach and Palm Beach, and in 1955 the city acquired the nineteen square miles that still remained. In 1964 the State of Florida conferred special protection on it. Butcher said the city values and protects the wetland for its importance to wildlife as well as its water supply function.

Today, Grassy Waters Preserve supplies water to the city's 130,000 residents. Butcher said that every time someone in the city turns on their tap, the water that comes out is from the preserve. I had never heard of such a direct connection between water usage and a local wetland.

I got a strange feeling looking down into the marsh because the two and a half feet of water was so clear it was nearly invisible. The play of light with the water had a surreal quality. Butcher explained that the water was so

FIGURE 9.1. On a guided kayak tour of Grassy Waters Preserve, Kait Kennedy admires the periphyton below the water surface from her kayak. Photo by Hilary Flower.

transparent because it is all from rain rather than runoff. And they keep it clean by restricting access and activities. Other than these regular guided tours, the public does not get to see much of the interior of the preserve.

I was peering through the water hoping to see the base of the snail kite food chain, the thing that the native snail depends on. And there it was. A buff-colored material, shining gold in the sunlight. It's called periphyton, and it is a collection of hundreds of types of microalgae, cyanobacteria, diatoms, minuscule animals, and microbes. Periphyton grows around ("peri") plant stems ("phyton") like a corndog. Sometimes it forms a mat at the marsh bottom.

Periphyton is not the green slime algae everyone hates: It is springy and full of water like a sponge. It has a microscopic structure made of calcium carbonate, the same substance that makes up the hard part of apple snail shells. I grabbed a blob of periphyton and squeezed it underwater. As it shrank in my hand, a glittering trail of bubbles rose to the surface. Oxygen bubbles! As the algae in it photosynthesizes, it produces a lot of oxygen like the bubbler in an aquarium. Calcareous periphyton is the favorite food of the Everglades' small grazers, including small fish, frogs and tadpoles, aquatic insects, rams-horn snails, and native apple snails. In the Everglades more than half of the photosynthesis is done by periphyton, so periphyton, not plants, is the foundation of the food chain. Our native apple snail specializes in eating periphyton, although it can eat plants.

After a short paddle, Butcher had all of us arrange our kayaks in a strip shaded by cypress trees so she could talk about what we were seeing under the water. I had invited Kait on this adventure because of a trip to the Everglades with her and another student of mine, Zoe Sabadish. While the other students slept in, Kait and Zoe joined me for a dawn swamp walk. At one point, feeling hot, I sat down in the cool water to dunk my head, and they followed suit. Zoe floated on her back, arms outstretched. Looking at the two of them soaking wet and beaming, I knew they were kindred spirits. Now, whenever I lead a swamp walk with students, I invite all who are interested to get fully immersed with me in the clear, cool water. Kait had done a class presentation on snail kites, so she was excited when she found a living native apple snail. She had raised her palm to show me, like it was a precious jewel, dark brown and about an inch in diameter, with its operculum closed up.

As we listened to Butcher share about Grassy Waters, Kait discreetly pointed her paddle toward a string of white pearls on a little cypress sapling. I looked around and pointed to another one, and with a smile she pointed to another. About a dozen native snail egg clutches were visible from where we were. I noticed that one egg cluster, on a skinny little cypress sapling, was still very pale pink, which meant that these eggs had been laid in the last day or so. They whiten as the calcium carbonate hardens. It was April, toward the end of peak egg-laying season for the native snails, but in Grassy Waters, they were still in full swing. By 2010 the bright-pink egg clutches of the invasive snail had started to become ubiquitous in many wetlands, and the pearly white eggs of the native snail had become ever scarcer. That day at Grassy Waters, I would have been overjoyed to see a single egg clutch of the native snails. Seeing dozens in one spot left me a bit disoriented.

Grassy Waters really was like going back in time. In the preserve, the original relationship between the kite, the native snail and the Everglades was still in place, as if all the rest had been a bad dream.

It was encouraging to see that the native snail was still thriving somewhere in Florida, even though the preserve was just 1 percent of the size of the Everglades National Park. Apparently, Grassy Waters held the recipe for happy native apple snails. I wondered if it had something to do with the glorious, healthy periphyton. I had come to Grassy Waters to immerse myself in one of the crucial wetlands that Jean Takekawa had fought to protect. I had not expected it to hold the secret to bringing the native snail—and the snail kite—back to the Everglades.

As the kayak trail started to bring us back toward the visitor center, I was sad to step back onto dry land. Kait and I lingered on the deck behind the nature center overlooking a lily-covered marsh. Kait had met Takekawa on one of the video interviews, and I thought about sharing the elder biologist's words of encouragement with her. But I realized I didn't need to. Kait already embodied the indomitable spirit of Takekawa. She would not give up.

Lauren Butcher joined us on the deck, and I told her how taken aback I was to see abundant native apple snail egg clutches and no invasive snail eggs. She said that she does see the pink egg clutches of the invasive snail along the edges, like near the parking lot, the boardwalks, and perimeter trails at the outer edges of the preserve. In the interior of the preserve, though, she mostly sees white eggs. With a big smile, she added that she usually saw at least one kite every day off that deck or through her office window. She loves pointing out the federally endangered bird to visitors, and watching the excitement of an experienced birder glimpsing one for the first time, making it one of the birds on their "life list." It was a snail kite sanctuary in all of the ways that it mattered.

Butcher told us about a time when she was kayaking out in a remote part of the preserve with a couple of colleagues. She said, "You come around a cypress dome and it opens up into this lovely almost secluded area. I looked over my shoulder, and maybe a boat length behind me, there was an adult male snail kite, right there with us, coursing over the water. He did not pay us any mind whatsoever. It was such a privilege for us to be fully in this bird's element, to get that intimate view into his world."

She rubbed her arms. "It still gives me goosebumps to think about. This moment out of time. It's transporting."

On the drive home Kait talked about her future plans. I asked whether she felt hopeful for wildlife, and for our own future on the planet. After a while she answered, invoking a term from wetland class: An "ecotone" is a transition zone between two habitat types. For instance, snail kites seek out ecotones for nesting: They need trees or tall cattail for their nests, right next to open marsh for foraging. Kait said, "I have always found the ecotone of hope and despair to be a convoluted affair. At the end of the day both come from love. Despair for what we've lost—the Wild Florida that lives on in stories and sacred spots—and hope for the future—the Florida we wish to pass to generations to come." She smiled, and added, "I guess hope really is the thing with feathers."

I said, "What do you mean?"

She was surprised I didn't know the poem by Emily Dickinson. She looked it up on her phone and read it to me as we headed west into the sunset.

"Hope" is the thing with feathers—
That perches in the soul—
And sings the tune without the words—
And never stops—at all—

And sweetest—in the Gale—is heard—
And sore must be the storm—
That could abash the little Bird
That kept so many warm—

I've heard it in the chillest land—
And on the strangest Sea—
Yet—never—in Extremity,
It asked a crumb—of me.

# PERIPHYTON WONDERLAND WITH NOBODY HOME

# 10

If Grassy Waters Preserve was a vision of the intact Everglades, Indian Prairie showed me what that same landscape looked like with one glaring difference: no snails. And no snail kites.

It was like arriving at a friend's house, the front door wide open, but they are not home.

Indian Prairie is a marsh at the northwestern edge of Lake Okeechobee, north of Moonshine Bay (see figure 1.2). I looked around, but I could not spot a single pearly-white egg clutch and no snail kites. As recently as 1995, there would have been plenty of both; back then, Lake Okeechobee was the second-most important snail kite breeding site year after year, second only to Water Conservation Area (WCA) 3A to the south.

With a pang, I looked over to Dr. Paul Gray, the Everglades science coordinator for Audubon Florida. Whole documentaries are devoted to watching him airboat around in the Everglades explaining everything you need to know about the snail kite. Earlier that morning, as he had emerged from his truck with bare feet, he looked like someone who might live on the water. Soon two graduate students from Florida International University (FIU) arrived at the boat ramp, Hanna Innocent and Tommy Shannon. Gray had towed behind his truck one of Audubon Florida's airboats that morning in order to take the graduate students out to their periphyton research sites. For my part, I was there on a mission to find out why the native snail had declined here and across the Everglades, and what could be done to turn that around.

Once we settled in the airboat with our protective headphones on, Gray turned the key, and a loud roar erupted from the aircraft-style engine behind our seats, enclosed by a big metal cage. Airboats' flat-bottomed hulls allow them to skim across shallow marshes into places inaccessible to traditional boats. Gray pressed the gas pedal with his bare foot, and we took off. He clutched what looked like a long, upright pipe—a rudder stick for steering. The two students sat in front.

Hanna had previously set up glass plates tethered to PVC poles at various locations throughout the marsh. Her advisor, Dr. Evelyn Gaiser at FIU, is the preeminent expert on periphyton in the Everglades. We stopped at each station so the students could sample the water and measure the periphyton growing on the glass plates. They would then compare the differences in periphyton across the marsh.

To my surprise, we passed through an area chock-full of snail kites. I had never seen so many snail kites foraging in one place. I lost count at sixteen; it was too hard to know which swirling kite I had already counted. Gray said into his mic, "This place is lousy with snail kites!"

But the raptors thinned out and disappeared as we approached the marsh known as Indian Prairie. It was one of the most pristine parts of Lake Okeechobee's northwestern marsh.

Gray stopped the airboat. Hanna collected samples and we motored on to the next field site. At the fourth and final one, Tommy Shannon fully immersed himself in the water, even though it was only twenty-six inches deep. He was visible as a snorkel sticking up between the blades of spikerush. He had an underwater camera that could take microscopic images so he could capture the tiny members of the periphyton "community."

I sat on the edge of the airboat and slowly slid off as my feet felt for the bottom. I sank the soles of my feet into the lovely periphyton mat. It felt like a thick, flexible carpet. Looking around, I realized that I had never been somewhere so flat. There were fewer shrubs and trees here than in Grassy Waters. A single continuous surface of water stretched out in every direction. The blue sky and puffy white clouds were perfectly reflected between the sparse vegetation. It was as if I were standing in the sky itself, if it weren't for the sparse vegetation and the occasional tuft of a wax myrtle bush or willow. I was surrounded by skinny little green blades of spikerush poking up a foot or so above the water. In the distance I could see patches of white lilies among lily pads so flat they appeared as disks etched into the surface. Dragonflies flew all around us. Once again I noticed the strange quality of light created by perfectly transparent water. The sunlight set the golden periphyton aglow. I started checking the periphyton sweaters

FIGURE 10.1. Dr. Paul Gray from Audubon Florida wades amid sparse and very skinny stems of spikerush, with periphyton visible below the surface, in Indian Prairie in the northwestern edge of Lake Okeechobee. Hanna Innocent can be seen conducting fieldwork on the airboat in the background as part of her graduate work at Florida International University. Photo by Hilary Flower.

on the spikerush stems below the water surface, hoping to spot an apple snail to take home with me.

Gray waded over to me. Given the lack of native apple snail egg clutches, this seemed like the ideal place for me to ask about the native snail's decline across so much of the snail kites' habitat. Paul Gray knew this marsh better than just about anyone, so I knew he'd have pondered this question.

First, I asked Gray to explain how Lake Okeechobee, notoriously polluted by canals, could host a periphyton wonderland like this. He explained that the canals do not open out on the edge of the lake; instead, they continue with fortified edges until they reach the deeper part of the lake. As a result, rainfall is the main source of water for the shallow marsh. That was another connection with Grassy Waters and the historical Everglades, I thought. Being rainfed made the historical Everglades nearly nutrient-free. Over the millennia, the periphyton communities and plants developed adaptations to survive. Saw grass, for instance, grows slowly, hangs onto whatever nutrients that do come its way, and uses nutrients extremely efficiently. Non-native plants have difficulty getting established under such austere conditions, allowing native plants to dominate.

In the presence of excess nutrients, though, the competitive advantage reverses. At first, the native plants respond like any plant would, growing taller and broader. But before long, exotic plants outcompete them and take over. Hydrilla, water hyacinth, and cattail can grow so densely that they block the light and squeeze out other plants.

Dr. Evelyn Gaiser's earlier long-term experiments revealed that even minimal doses of phosphorus in an Everglades marsh led to a complete collapse of the calcareous periphyton and native plants. Within five years, it was nothing but dense cattail, a condition known as a monoculture. This is bad news for native snails and the rest of the Everglades.

Gray asked if I knew how to read the nutrient levels by looking at the vegetation. I couldn't wait to learn; I love clues for reading invisible things in the environment. He reached out to touch one of the needlelike blades of spikerush and turned to me with a smile, saying, "Notice how skinny these are. The water is so pure here, the blades are thin. Later, as we go back toward the more nutrient-enriched areas, you'll see the blades get thicker and thicker." He pointed to a patch of sagittaria, and asked, "Have you ever seen sagittaria leaves that narrow?"

I had not. On the college campus where I work, the retention ponds are ringed by sagittaria, and they are robust affairs, with lance-shaped leaves as broad as my hand. Here, their three-petaled white flowers were smaller, and their leaves were short, sparse, and narrow as grass. Now that Gray had pointed it out, I saw the low-nutrient version of familiar plants all around me. The blue sky reflected between the spikerush blades created a pattern of blue and green vertical lines shifting and almost vibrating as I moved.

A snail kite flew over us and landed on a little wax myrtle bush fifty yards away. I looked around and realized it was the first snail kite I'd seen in this relatively pristine area.

I said to Gray, "This place is perfect for snail kites. So, where are they?"

"It's because the native snails are not doing well here," he explained. "They're not doing well anywhere," he said, with a shake of his head.

Gray said that by 2010, native snail populations had started to dwindle on the lake. If the snail kites were still relying on the native snail, they would probably be extinct by now. This was no longer their home.

Gray pointed out what looked like a large wad of pink bubble gum on a stalk of vegetation. It was the unmistakable egg clutch of the invasive snail. But that was the only one we found there. We were in a virtually snail-free marsh. Gray said this was not the preferred habitat for the invasive snails: It was heavy on the periphyton, light on the plants, the opposite of Paynes Prairie.

The snail kite flew away, and I watched it become a dot and disappear. There was nothing for it here.

I had previously wondered if the native snail population crash had been caused by a dwindling of periphyton, but it didn't look like it. "It sure looks like the periphyton is doing well here," I said.

Gray agreed and pointed out that the graduate students' work that morning had provided the perfect demonstration. Where the water was dirty, the algae on their glass slides was the thin green kind. But out at Indian Prairie, where the water was clear and clean, we saw a thick, buff-colored layer of the calcareous periphyton on the slides.

I told Gray that I was baffled that snail kites could be missing from a marsh that looked so perfect for them. Yet there were too many snail kites to count in the more impacted parts of the lake we had passed through earlier in the day.

Gray considered. He said, "We have designated 'indicator species' in Everglades Restoration because, if these certain organisms are doing well, we can tell that the ecosystem is working. And they tend to be apex predators, like wading birds. . . . You know, if the wading birds aren't doing well, it's either because the water is too deep, or there aren't enough fish, or grass shrimp, or crabs, or whatever. Snail kites are the same way. If we're managing the water right, and the native snail's doing well, then the snail kite numbers will be good. It's a Goldilocks species for how we're managing the Everglades. And how we're managing Lake Okeechobee."

I liked how he explained that, the same linear cause-and-effect chain I taught my wetland students. I found it both fascinating and humbling that animals provide more reliable indications of ecosystem health than the individual components that we can directly measure, like water quality and depth. Ecology is still a young science, and ecosystems are complex.

Until the last couple of decades, Gray said, the snail kite was tightly tied to the native snail and its water level specifications. Thus, snail kite numbers faithfully reflected how the ecosystem as a whole was functioning. I remembered that back in 2012, the National Research Council evaluated Everglades restoration progress. They graded restoration based on ten critical ecosystem features, and snail kites were one. Each got a letter grade like in school, where F meant "failing," which is to say, near irreversible degradation. Restoration got an F based on the dire and possibly irreversible degradation to snail kites' habitat, which so many other species relied on as well. The report noted, "Because Florida snail kites are considered to be near extirpation, loss of even a single kite or kite nest is a serious matter." Gray remembered that infamous F well, and he agreed with the assessment.

The invasive snail changed everything. Snail kite numbers have recovered to the 3,000 level that we saw in the late 1990s, but not because the Everglades habitat improved. Snail kites abound in wetlands with abundant invasive snails and the invasive plants the snails love to graze on.

Gray said, "Right now, there are no native snails in Lake Okeechobee. Well, 'very few' is more accurate. But the snail kites are nesting in Moonshine Bay because of the invasive snail, which doesn't need a healthy wetland. Their indicator status has gone awry."

I understood what he meant. I figured that if we could get the Everglades water management right, the native snail population would rebound. I was curious to hear what Gray thought the pathway forward was for putting that into action. But before I could ask, he said, "Right now, the native snails are not doing well in the Everglades. And," he said, "we don't know why."

Wait. What?

This was the first I heard of that. I assumed that scientists knew *why.* Indeed, I had assumed that it was clear what needed to be done and that only the political will was missing. Water always seems to be tied up in politics, especially in Florida.

Gray made his point even more clearly: "Here we have the kite whose population dropped by 75 percent in ten years, and now it's rebounded back up to precrash numbers. But it's not because of the Everglades. And it's not because of native snails." He paused. "We don't know *why* the native snails won't come back in the Everglades. *That's* the big problem."

I was dumbfounded that he was dumbfounded.

I said, "It seems like the typical suggestion is that water management is not getting the water right. But it sounds like we do not know *what* to change now, to make the Everglades better for them?"

"Yeah," he said, putting his hands up. "We don't know!"

Gray urged me to reach out to Florida's apple snail expert, Dr. Phil Darby. If anyone knew what the problem was, it would be him.

Shannon resurfaced from snorkeling, beaming as he held up a surprise: an apple snail. He admired it before wading over to pass it to me, having heard me say that I'd been searching for one.

I was delighted to hold the plump snail in my hand. The water clinging to it caught the sunlight. The fine concentric ridges of its operculum shone like polished wood grain. The shell had prominent brown and gold stripes. I inspected its pointy top and observed the deep grooves along each turn of the spiral, the telltale sign that an apple snail is the invasive one and not the native (see figure 2.2).

As I looked north over Shannon's shoulder, I saw a distant bank of storm clouds, with a thick, gray shaft of rain angled down. I could feel the sun on my arms, and I wondered if the storm would come our way. I was a little concerned about us being the only high points in a thunderstorm. Still, summer storms in South Florida can be remarkably isolated, and boaters can sometimes navigate to avoid them.

Hanna said she was almost done with her sample collections.

As we climbed back onto the airboat, Gray shared two troubling facts about the invasive snail I held in my hand. First, the snail kites were relying entirely on the invasive snail. Entirely. And second, he said, the invasive snail may not be a reliable substitute. Gray explained, "The invasive snail seems to be a disturbance species." I had heard that expression before. I knew that in ecology, the term "disturbance" covers any change in environmental conditions, such as a hurricane or a wildfire, which is big enough to change how the ecosystem functions, at least temporarily. I wondered what kind of "environmental disturbances" the snails responded to. I added that to my mental list of questions to pursue.

Gray went on, "It's a boom-and-bust thing. It comes to an area. It'll reach fantastic densities. And then it'll drop off. The next year, there will be hardly any of them! So, the kites find them, and then track them as they pop up in new places. What's really sustained the recent population growth of the snail kites is these nesting supersites that happen every year somewhere, based on a boom of invasive snails. And we'll have two hundred nests, or three hundred nests, in one spot. And we'll have a bunch of babies. But if you take away these boom sites, the sites that are there one year and not there the next, the kite population would not be increasing."

Oh, no, I thought.

Gray said, "We're relying on a random event. We hope somewhere this year something weird will happen. Relying on random events to sustain your population is not the most secure place to be."

I didn't know what to say. I ventured, "That really scares me."

Gray nodded with a somber smile. "Well, not as much as it scares us. We're reliant on something that's kind of unreliable. . . . And unpredictable."

The other shoe had dropped.

To drive the points home, he raised a finger to indicate the first item in a list: "Not only are they not nesting in the Everglades and, you know, for the Everglade snail kite not to be nesting in the Everglades. . . . It's not right."

Then he raised a second finger with emphasis, saying, "But they're also relying on a species of snail that's not native. And we don't know enough about them yet to know if we could trust them, so to speak."

Without noticing, I had come to think of the native snail as a "nice to have" but not a "have to have" for snail kites going forward. I took for granted that kites could rely on the invasive snail, and my main question was whether they could ever return to the Everglades. I did not realize that the ordinary densities of invasive snails were not enough, that a population explosion was necessary for the invasive snail to support large-scale snail kite nesting. That had not been the case for the native snail.

With a stab of recognition, I thought of Paynes Prairie. While I was marveling at the numerous snail kites there with Dr. Caroline Poli, I was unwittingly bearing witness to the tail end of a boom-and-bust cycle, as the kites took up to an hour to find a snail. Indeed, kite nesting at Paynes Prairie turned out to be a lot lower than expected that year.

Gray pointed out that the native snail is the snail adapted to live here. Hence, it would be the species most reliable to carry the snail kites into their future. That made sense. It made it all the more urgent to figure out what was going wrong for the native snail across South Florida, and to rally the troops to address the problems. All in the narrow window of time before we have a string of years without big snail booms.

We got sprinkled on a bit, but the storm did not hit us. As we motored away, I looked over the edge of the airboat at the marsh speeding away. Before my eyes, the periphyton wonderland with sparse spikerush gradually disappeared. Gray pointed as the spikerush stems became thicker, taller, and closer together. We were heading along the nutrient gradient he told me about, as we left the clear water and approached the more nutrient-enriched areas. The grass-thin sagittaria became taller and thicker until it resembled kayak paddles along the canal. Along the gradient of increasingly poor water quality, snail kites went from nonexistent in the clean water to uncountably abundant in the impacted area, happily foraging on the invasive snails. It was a visual enactment of how the snail kites' indicator status had gone awry.

I realized that the kites' new preference for impacted wetlands would turn our conservation strategies upside down. Would we have to embrace excess nutrients and invasive plants? Would we have to wreck our lakes to save the snail kite?

I resolved to get an interview if at all possible with apple snail expert Dr. Phil Darby. In the meantime, I would learn everything I could about the invasive snail. Thus far I only knew how to identify it and that it preferred plants over periphyton. That also explained why it was only on the periphery of Grassy Waters Preserve. I was curious what else differentiated the two species of snail. I could not guess what might cause the boom-and-bust cycles that Gray talked about. I realized I didn't even know when the invasive snail had started taking over the state. I couldn't help but wonder if they were to blame for the scarcity of the native snail.

Once again, I saw that when it came down to it, this quest to understand the plight of snail kites was a tale of two snails, and two very different versions of Florida. The native snail represented the healthy ecosystems of predevelopment Florida, still very much alive in rare places like Grassy Waters. The invasive snail represented the infusion of nutrients and invasive species that have been compounding over the decades, especially in Lake Okeechobee and the lakes of human-dominated Central Florida. Which snail, and which version of Florida, would best carry the snail kites into the future?

As we flew through the marsh, I felt the fat little invasive snail in my hand. It struck me that I was holding the snail kites' present. But was it also the key to their future? Could this invasive snail save this endangered species? Or were they a source of false security, giving the kites a last hoorah before extinction?

# THE SNAIL IN MY KITCHEN

# 11

I was excited to bring a snail home from Lake Okeechobee, because getting to know this important character in the snail kites' story had become my top priority. So ubiquitous, and yet so elusive. From pictures online I couldn't even get a clear sense of what their bodies looked like or how they moved. I was also eager to dig into the scientific literature to answer my questions: Where did it come from? How did it get here? Why was it labeled "invasive"?

I didn't have an aquarium yet, so for the first night I put it in a bowl with water and a leaf from a weed in my yard. I knew apple snails could breathe in air and underwater, so I figured it would enjoy its snack and get some sleep.

The next morning, it was gone. The bowl had no leaf and no snail. My snail had escaped!

After some frantic searching, I found it under the kitchen table with a small chip in the edge of its shell from its fall. I decided to name it Chip.

The scientific name for this species is *Pomacea maculata,* and the *maculata* part refers to spots or speckles, so they are "spotty apples." I had read that the spots were on the inside of the shell, but that was never the case in the empty shells I found. The first time Chip came out of its shell, I had the answer to that mystery at least: It had tiny freckles all over its beige, fleshy body.

This snail species is native to South America. Its common name, which no one in Florida seems to use, is the giant snail; it can grow up to five times larger than Florida's native snail. The novelty of its large shell appeals to people with aquariums. It is one of a few different apple snail species which are sold in pet stores under the name "mystery snails." Typically, Mystery Snails are *Pomacea bridgesii,* the kind Jean Takekawa found in a farm canal in 1982. Sometimes owners get tired of having them, so they dump them in a wetland, not realizing how damaging this can be to ecosystems. When exotic species are introduced to an unfamiliar environment, many of these newcomers die or develop a small population—but some take off. It amused me to think that by bringing an invasive snail from a wetland into an aquarium, I was reversing the journey of its forebears.

Invasive species often share a set of characteristics, of which *P. maculata* is a shining example. They typically reproduce rapidly. Our invasive snail can mate early in life: as young as three to four months of age, compared to about a year for our native snail. The invasive snail puts out egg clutches more frequently during the usual breeding season, and each clutch has thousands of tiny eggs, compared to the tens of eggs in native snail clutches. The invasive snail also puts out more egg clutches during the off-season than the native snails do.

I remembered Dr. Caroline Poli, when we were out at Paynes Prairie, describing going out in airboats for snail kite surveys, and they'd have to wipe the "pink and goopy snail eggs" off the side of the airboat. She had gestured with her hands in a wiping motion and made a yuck-face.

Around 2011, on a ranger-led canoe trip with my children and a dozen students, the ranger pointed out the eggs of the invasive snail and encouraged us to whack them into the water with our paddles. We were all scandalized to have a green light for destroying wildlife. Many pink eggs got whacked that morning.

Another characteristic shared by many invasive species is being generalists. They can chow down on whatever food is available in new areas. The invasive snail has a particular taste for invasive plants, making contemporary Florida a very welcoming place.

The invasive snail is also flexible about water depth; it loves plants like cattail that thrive in deeper water. And it can handle the other extreme better, too; it is more capable of surviving dry downs than the native snail. Apple snails can go dormant in the dry season by tucking under the muck and closing their operculum, sealing in moisture. The invasive snail can survive in this state for longer than the native one; being a larger snail means you have greater reserves of both energy and moisture for a longer period of dormancy, also known as aestivation.

The invasive snail can handle more extremes of temperature. Tommy Shannon, the FIU graduate student who had handed me the snail, told me he had found snails in water that was 105-degrees Fahrenheit. They are also more cold-tolerant, allowing them to press northward beyond Florida into several other states. Their superior ability to aestivate also helps them survive cold temperatures, closed up under an insulating layer of muck.

The most important difference is that the invasive snail lives up to three years, and sometimes more, compared to a year for the natives. Three years of making babies means a lot more reproductive potential.

All these factors have helped the invasive snail to spread broadly in Florida, and beyond, into a few other states in the US. Similarly, it has spread around the world, brought in either for food or the aquarium trade. And in most places, it's unwelcome. It made the list of the one hundred "World's Worst" invasive species, compiled by the Global Invasive Species Database. I imagine this snail's tentacled mug emblazoned on a "Most Wanted" poster.

But why? What harm had it done?

The term "invasive" is bestowed by government agencies on non-native organisms that have damaged the environment, human health, or the economy. If a non-native species doesn't take over or do harm, it doesn't qualify as invasive. When a non-native species does become overabundant, it's often because its new environment lacks the ecological controls that kept it in balance in its original home. Cattail is a native plant in Florida, but the extremely low nutrients in predevelopment Florida prevented it from spreading into the monocultures we see today.

This voracious snail has earned its "invasive" label principally due to its devastating effect on rice paddies in Europe, Asia, and South America. Because of this global reputation, when it began to spread across Florida, the "invasive" label seemed a natural fit. However, there's not much rice farming for it to ruin in Florida. As for natural wetlands, I could not find reports of them damaging any large natural wetlands in Florida or anywhere else.

Eyeing Chip in its bowl, I realized I felt repulsed. If it had been a native snail, I think I would have found it charming and endearing. Even knowing this, I couldn't shake the feeling. I could only assume that the label "invasive" had engendered an adversarial posture in me, a desire for Florida to be free of it. I talked about this with Sal, now twenty-seven years old and working with invasive plants in Virginia. Sal pointed out that, as with the term "illegal alien," the way we talk about invasive species often has xenophobic connotations. Demonizing "the other" can affect how we treat people from other places as well. Also, by labeling a plant or animal invasive, and using other emotionally laden words like "infestation," we shift the blame away from us, even though it's our actions that brought them here. They never asked to be moved; they are just organisms living their lives. Indeed, many invasive species are imperiled in their regions of origin. Burmese pythons are invasive in the Everglades and considered vulnerable in their native ecosystem in Southeast Asia.

To be clear, this is an argument for being thoughtful in our terminology, not an argument for ignoring them. Invasive species can devastate ecosystems; Burmese pythons have caused a 99 percent decrease in mammals like raccoons, opossums, and rabbits in the Everglades in just two decades. Florida is overrun with reptiles, fish, and plants like water hyacinth that came to the state as part of the pet or aquarium trade. Our tropical environment makes it easy for a lot of exotic tropical species to thrive.

Another reason for me to look kindly on Chip was that its species had single-handedly saved the snail kites, at least for now. The limpkin had also rebounded due to the invasive snail. This reminded me of a dilemma I had read about in the Western US. Managers were trying to eradicate an invasive plant called tamarisk, but they had to stop when they discovered that an endangered bird known as the southwestern willow flycatcher relied on it for nesting. Its native nesting material, the willow that gives it its name, was no longer available. I was fascinated that the snail kite story was one of these rare instances where an invasive species is saving an endangered one. Until the native snail recovers, we may do well to embrace this particular non-native snail in Florida.

Still, I figured there must be some ecological downside for having the snail proliferate so much in our state. One study pointed to its destructive *potential* to Florida's wetlands. The researchers compared the eating habits of the native snail vs. the invasive snail in tanks with tape grass. The native snail was a dainty eater: It ate along the edges of the green blades, causing minimal damage and even spurring plant growth. The invasive snail was a comically messy eater: It would chop through the middle of stems, or mow through the plant's base, leaving the rest to fall as detritus. In this way, the invasive snails killed large chunks of plants and stunted plant growth. They consumed plants ten times faster than the native snail. And when

FIGURE 11.1. An invasive snail from Lake Okeechobee, whose shell got a chip upon being brought to the author's house, and then grew rapidly in the following weeks. Illustration by Hilary Flower.

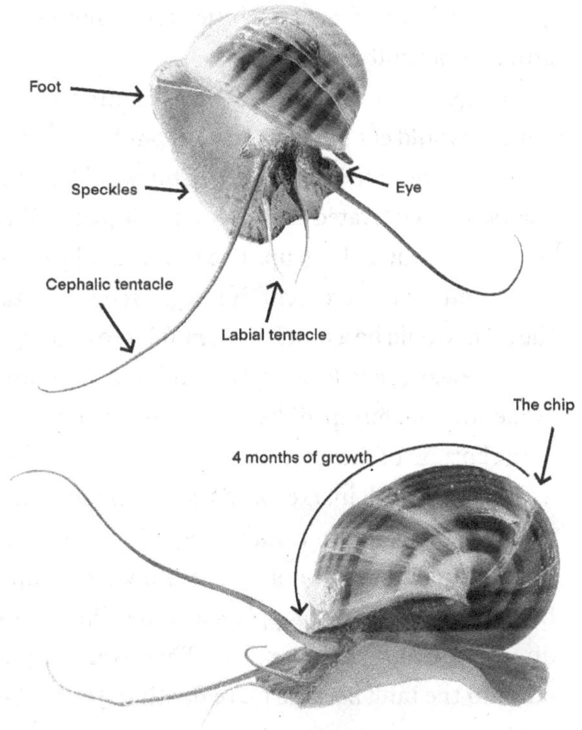


they chopped through the vegetation, it would rot, raising the nutrient levels in the tanks.

Dead vegetation. Murky water. Putrid odors. I should have thought twice before adopting one.

The day after I got Chip, my neighbor's teenage son, Henry Rumschlag, came over with a loaner tank and some bottles to get the water chemistry right. He told me his mystery snail keeps its tank clean by eating algae from the glass. I had seen the beautiful aquatic plants Henry was growing in his own tanks, and he promised to help me do the same.

When the tank was full of clean water and there was a nice leafy sprig of vegetation in there, I gently dropped Chip in. Chip slowly sank to the bottom. Henry and I leaned down to watch as it gradually opened its operculum, which was attached to its fleshy foot. When the operculum was pretty far extended, it tipped to the side to allow the operculum to flip upside down behind it. It crawled forward with the operculum face up in its wake. It had eyes on little stalks, and long tentacles that it waved around gracefully.

Things went downhill as soon as Henry left.

Chip would eat leaves as fast as I put them in the tank, leaving only pale stems floating at the top. Clearly, Chip would not be cleaning algae from the glass while I cultivated attractive aquatic plants. Weeds from my yard would be the only way to keep up. The water rapidly became murky from rotting stems and snail excrement. If I neglected Chip's tank for more than a day, the stink would be a powerful reminder to empty, scrub, and refill the tank.

My reward for cleaning the tank was watching Chip motor around. Sometimes if I bumped the tank or otherwise startled Chip, it would drop to the bottom of the tank and withdraw into its shell, closing its operculum.

About a month in, I realized that Chip's shell had grown an inch beyond the chip it had gotten from falling off my kitchen table. The next month, another inch. I could not believe how fast Chip bulked up. It had gone from golf-ball-sized to apple-sized, heading in a distinctly softball-sized direction. Chip was heavy, too. That explained how quickly the plants I put into the tank became mere floating stems: Chip was packing it on.

More leaves. More murky water. Day in and day out, rapidly enlarging this snail.

I started to buckle under the burden of my snail overlord.

No offense to Chip, but I did start to realize how someone who bought this species under the name mystery snail could consider their options for off-loading it. Even though I knew of nearby wetlands where this species could be found, it is deeply ingrained in me to never release a non-native species into the wild. Some of Florida's Burmese pythons are thought to have been released into the Everglades by disaffected pet owners.

I offered my snail to a nearby raptor rehab center, but they worried that it could have parasites. Eventually I found a student who already had a *P. maculata* for a pet. So, I rationalized, he knew what he was getting into, and he was fine with it. I brought Chip to my office, and the student put Chip in his pocket and headed to his dorm. I was glad to have my old life back.

I was eager to apply what I had learned to a real wetland. If one snail could do this to a tank, what could an army of them do to a wetland? And since snail kite nesting supersites have centered around invasive snail population booms, I was eager to understand what conditions would light the fuse for these booms. Dr. Paul Gray of Audubon Florida told me about a stormwater treatment area where the snails had fully taken over and defoliated the plants, causing them to shut it down. That seemed like my ticket; I wanted to know everything.

# HOW TO MAKE A SNAIL BOOM

# 12

For Eric Crawford, the Snail-pocalypse started with a phone call. A coworker called to say, "I'm seeing a lot of egg masses. It's getting thick now."

When I interviewed him by phone, Crawford had a quick wit and a ready laugh. He was the vegetation management scientist for the Stormwater Treatment Areas (STAs; see figure 1.2 for location) for the South Florida Water Management District (SFWMD).

The call about the egg masses came on August 6, 2013, a day he'll never forget. The invasive snail had been established in all the STAs for a few years, so the bright-pink egg clutch were a common enough sight by then. Crawford had vaguely noticed them becoming especially prominent in that particular wetland cell. Now he wondered, Could it really be that much worse all of a sudden? He had to see for himself.

He got in his work truck, and in a couple of minutes he was standing on a levee overlooking a marsh he had planted. He had never seen anything like this before or since. It was pink. Everywhere.

Each stalk of cattail had its own pink wad, and some leaned under the weight of several pink wads. There seemed to be as much pink as there was green. Even the yellow vinyl flow barrier that snaked across to the wetland was festooned with pink polka dots.

Crawford walked down from the levee to get a closer look, letting the water sink into his field sneakers. A large snail laid its eggs on a cattail stalk. As Crawford scanned the cattails around him, he saw more snails laying eggs. Who needs the cloak of darkness when you have the safety of a crowd? He watched a snail lay eggs on the yellow flow barrier—in broad daylight. Like it owned the place.

Crawford realized that the invasive snail population must have been slowly ramping up for a while without anyone particularly noticing. Now it was impossible to miss.

FIGURE 12.1. The invasive apple snails that Eric Crawford of the South Florida Water Management District collected in just a few minutes on August 6, 2013, the first day he became aware of the population explosion in a Stormwater Treatment Area southeast of Lake Okeechobee (specifically, cell 4-south of STA-1 East; location shown in figure 1.2). Photo by Tony Griffin.

It was taking place in a seven-hundred-acre cell in one of Florida's human-made wetlands known as Stormwater Treatment Areas, or STAs. Just south of Lake Okeechobee is the Everglades Agricultural Area (EAA), where sugarcane and other crops are grown. Phosphorus is a common fertilizer used to boost crop yields. Nutrient-enriched runoff from the farms must be cleaned before it enters the Water Conservation Areas (WCAs). So, the STAs were built between the EAA and the WCAs (see figure 1.2). They might as well be called *Farm*water Treatment Areas.

I love that the most cost-effective way to remove phosphorus pollution from billions of gallons of water daily is not through chemical engineering but through good old-fashioned nature. As water flows very slowly through a wetland, the plants use the nutrients to help them grow, storing it in their tissues. When they die and become part of the soil, the phosphorus remains locked up in the soil. Each wetland cell is bounded by canals and levees. The water flows first through one vast cell, then into the next. After all that contact with wetland vegetation, the outflow water is low in phosphorus, and it is ready to flow into the WCAs and the Everglades.

Crawford told me he that as he tried to discover the extent of the problem, he had jammed his hands into the base of the cattails and started feeling around. He recalled, "I hadn't been in there five minutes, and I had a handful of snails. I was like, 'This is a thick population!'"

Next, Crawford dug up some cattails and looked at their root base. He said, "They were full of tiny little three- to four-millimeter snails hiding between the layers of the cattail leaves, eating everything." I imagine the camera gradually panning out from tiny snails on a cattail leaf to reveal an aerial view of Crawford surrounded by the equivalent of 530 football fields of pink-smeared cattail.

That was when he became truly frightened. This was a huge, unprecedented problem. It was his job to solve it. And he had no idea how.

To see this happen to this cell was especially painful, because this cell was his baby. When the STA had first been constructed in 2005, Crawford's predecessors had not been able to plant successfully in the outflow region. When Crawford arrived, he put in two years of trial and error and finally had a strong and functional outflow vegetation strip. Thanks to his diligent work with the plants, that STA removed about 80 percent of the phosphorus from the inflow water. It was his first big success.

All that progress was reversing before his eyes. This cell had submerged aquatic vegetation (i.e., plants that grow only under the surface of the water), interspersed with sturdy cattail and bulrush to protect them. All around him, dead vegetation was floating on top of the water. As vegetation rots or becomes snail excrement, nutrients that had been bound up in the plant tissue are released into the water.

Dead vegetation. Polluted water. I thought of Chip's tank stinking up my kitchen, but a square mile in size.

Back at the office Crawford braced himself and checked the phosphorus levels. It was worse than he'd feared. Remember how the job of each cell was to take up phosphorus, making the outflow water cleaner? The cell taken over by snails was now *producing* phosphorus: There was a lot more

FIGURE 12.2. Loads of invasive apple snail egg clutches are visible on cattails, where Eric has placed a square "quadrat" (a half meter on each side) for measuring the density of snails and their egg clutches, on August 6, 2013. Photo by Tony Griffin.

phosphorus in the outflow water than in the inflow. It was like a vacuum cleaner that suddenly started spewing dirt everywhere.

Crawford quickly asked the water managers to shut off the outflow. The next day, he took an airboat out into the middle of the cell. He cut the engine and slipped off the edge of the boat into the water. He dropped down a square PVC frame, known as a quadrat—a convenient way to mark off an exact area—so he could count the egg masses and snails and compare one area to another. The quadrat was a half meter on each side. He reached inside the square and started feeling through the muck and vegetation.

He told me that in the small square of his quadrat, he found 150 egg clutches, and sixty snails. He said, "Sixty snails! You can't even fit sixty snails in that! It was just snail after snail after snail."

Shaken, he hoped that spot was a fluke. He went to another spot and dropped the quadrat. He got similar numbers. He told me that he had done surveys for native snails in the past, and after three hours of "slogging and throwing quadrats," he'd be excited to get a single one. But now, as he put it, "You couldn't *not* get an invasive snail."

He sighed. "That was when I got depressed."

As his airboat passed slowly through the wetland, he noticed that plants were "rolling up" in the boat's wake. He explained, "The plants were no longer rooted. The snails had mowed them off the bottom. They had just been dying in the water column." In case there was any doubt as to the cause, empty snail shells were also rolling up in the wake behind the boat. He figured the empty shells were from bird and mammal predation.

"The worst part," he said, "it really stank."

There were large bubbles at the water's surface. It's not unusual for Crawford to see "good" bubbles in the STAs. In the summertime, he told me, photosynthesis is so intense he has seen oxygen bubbling off the bottom of lotus leaves in his tanks. But Crawford could tell by the pungent smells that these were bubbles of hydrogen sulfide, methane, and other gases from decomposition. Crawford said, "The water went from having high dissolved oxygen, typical for an STA cell, to being some anaerobic, smelly thing. It takes extreme conditions to force that."

All around him Crawford saw what looked like straw: hydrilla stems after snails had consumed all the leaves. It made me think of the pale stems I removed from Chip's tank. It turns out that hydrilla is one of the invasive snail's favorite foods. Crawford noticed that the snails were eating the hydrilla more than any other plant.

After Crawford's first shocking quadrat survey, the next week he saw even more snails in his quadrats. He saw yet more the following week. By September he was getting an average of 324 snails per square meter in open water areas. He estimated one million snails per acre.

He stopped doing quadrats, because, as he put it, "It was too depressing." He said, "It was just so fast. By the end of August, all of the plants were gone."

Gone. A bit more than a square mile. Stripped bare. Open water.

Crawford said that he knew that invasive species tend to be generalists, but he never could have imagined *all* the plants being eaten.

"It was really scary," Crawford told me.

I asked him, "Do you know what set off this population explosion?"

He couldn't say for sure, but he had a very convincing guess. It came down to the fish that normally keep snail populations down by eating their tiny hatchlings. Two factors shielded the snails from these fish and other predators. First, murky water. In the weeks leading up to the explosion, the water had turned to Ovaltine, as he put it. The US Army Corps of Engineers was redoing some structures in the upstream end of the cell, which created a plume of suspended sediment. That's why that big yellow barrier was in place: to keep the plume of cloudy water from spreading downstream. Any fish that relied on sight to hunt would have been unable to spot the snails. Same with sight-feeding wading birds like herons. Snail kites, too, would have been blind to the snails, even as the snails began their takeover.

The double whammy was that initially, hydrilla covered a large swath of the cell from one side to the other. Hydrilla grows thickly and can fill the water column, providing food and shelter for snails. Crawford suspected that the hydrilla made it harder for aquatic predators in two ways: blinding them to the snails and physically blocking them. He recalled, "The hydrilla was so thick, it would be hard to imagine a fish being able to swim through it. The water was a foot and a half to two feet deep, and I saw herons walking on it."

I interrupted. "Wait, really? Herons could *walk* on it?" This was like a whimsical cartoon makeover of a wetland.

Crawford laughed. "Yes! They were just walking on top of the water! That's how thick the hydrilla was. The baby snails were growing in *that*. It was the perfect environment for them to grow in."

Since invasive snail eggs are so tiny, their hatchlings are also tiny at first, with paper-thin shells. This makes them palatable to a wider range of predators than the adult snails. By having thousands of tiny eggs in a single egg clutch, the species is compensating for the extremely low survival rate.

The thick hydrilla drastically changed that equation. Crawford made some calculations based on population growth and hatchling survivorship values. He estimated that if conditions were to allow their usual survival rate to increase to just one in a thousand, that was enough to explain the population boom he had witnessed.

"It was a perfect storm," he said. "It's extremely rare for the hydrilla to be that thick in the STAs, and for it to fill the cell from bank to bank. And it's rare for the water in an STA to be murky. Between the two, the snails just took off."

Crawford reflected, "By the time we noticed, and we were like, 'Wow, this is weird, there's a lot of pink out here, and there are shells everywhere, and you can't walk around without stepping on a limpkin . . .' By that time, it was too late."

I thought about my conversation with Dr. Paul Gray of Audubon Florida out at Indian Prairie. He had lamented that we were relying on a random event each year to sustain the snail kite population. All the big snail kite supersites in recent years had involved explosions in snail populations. I was finally starting to grasp what caused the booms: circumstances that turn the predator dial way down. All the better if the invasive plant dial was turned way up. If you can suppress the invasive snails' natural predators and give them their favorite food, then boom! Snails everywhere.

And snail kites took full advantage. That STA had the second-highest number of snail kite nests in the state in 2013, second only to Lake Okeechobee. Crawford said, "The kites were already nesting there before the surge in snail explosion was apparent to us. The birds have much better eyes!"

It almost seemed like we could use this formula to proactively create nesting habitats for snail kites somewhere each year.

Now that I had a sense of what could lead to these population booms, I was eager to know why they always turned into a bust. Why did the snail kites keep needing to search out a new boom somewhere else?

Crawford could not afford to wait a few years for a bust to unfold; he needed to rapidly flip the switch from boom to bust. He explained to me how he did that when I met him in person out at the STA.

# 13 HOW TO BUST A BOOM

Eric Crawford emerged from a big white truck to greet me with a firm handshake. "SFWMD" was emblazoned on the truck, his gray baseball cap, and his gray polo shirt. The Stormwater Treatment Areas (STAs) are run by the South Florida Water Management District (SFWMD), often referred to simply as "the District."

He gestured for me to get into the passenger seat. "So," he said, turning to me with his hands on the steering wheel. "Do you want me to take you to where the Snail-pocalypse happened?"

"That's what I came here for!" I said.

When we got to the cell where it all went down, Crawford stopped the truck and pointed out his window at the endless cattail, useful for its ability to calm the water flow. We scanned the nearest stalks for pink egg clutches but could not find one.

Over the roar of AC coming out of the truck vents, Crawford told me that when he was grasping for a solution to the Snail-pocalypse, he kept thinking about a horror story from a fifteen-acre stormwater treatment pond at Martha Wellman Park in Tallahassee. I had read about this. In 2007, construction of the pond was completed, and the city put in a half million dollars' worth of native wetland plants. When they flooded it, the invasive snail wiped out three-quarters of these new plants in a single growing season. The city hired a firm that designed a snail trapping system. They removed four tons of the invasive snails in the first year. They also stocked the pond with native redear sunfish that love to eat the juvenile invasive snails. The pond recovered and got a new planting.

But traps and sunfish did not seem like a practical solution for a wetland fifty times larger.

I had read that rice farmers often use molluscicides, but they are expensive and can harm wildlife. At one rice farm in Ecuador, researchers investigated whether the resident snail kites could significantly curb snail populations. The answer was no, unless they installed artificial perches for the kites. In Crawford's STA cell, even at the ambitious consumption rate of sixty snails per day, it would take 16,000 snail kites a full year to eat *half* the snails in Crawford's cell. So even without installation of perches, snail kites were not the answer.

One of Crawford's options was drying the cell down. Invasive snails had overrun one of his small test cells the year before. He had dried it down, exposing the snails to avian predators, and also killing new hatchlings, which die if they don't fall into water. It worked. After slowly reflooding it with water, it never had another snail population surge. But again, Crawford worried about scale. I was drawn by the rugged experimentation-on-the-fly and educated guessing involved in Crawford's job, where real stakes were ever at the forefront.

Finally, Crawford concluded that a dry-down was his only hope. He put in the request to drain the cell. The water flowed away, exposing the wetland bottom. The soil stayed moist, because the canals ringing the cell retained water, and there's always some subsurface inflow from the canals. But it was enough.

Snails, tilapia, and other aquatic animals congregated in the canals and the shrinking pools of water on the mudflat. As if by invitation, the predators flocked in. Wading birds and gators gobbled up the highly concentrated prey.

Crawford said, "I think gator densities are higher in these STAs than anywhere in the world. When you drain a cell, somehow word gets out to the gators. They will migrate to canal banks and the exposed soils. They'll be all lined up with their sides touching. They waddle around after gorging themselves." He laughed and shook his head in disbelief, saying, "Same goes for birds like herons and egrets. You're measuring the bird density by the square meter, instead of by the acre."

I asked if he saw snail kites, given that the STA was a supersite for snail kite nesting that year.

He shrugged with a bemused expression. "A few snail kites would have easily disappeared in the thousands and thousands of wading birds foraging there."

Raccoons and even otters joined in the feeding frenzy. I did not know that otters ate apple snails, but Crawford told me it was normal for both otters and limpkins to make little piles of discarded snail shells at the STAs. But this time was different. He said, "The otters and the limpkins piled up shells by the thousands. When the water was down to the point where there weren't any fish left, we saw otters rolling around in there. It was funny watching them lurch across the mudflat."

Amid the circus-like atmosphere of the feasting gators, birds, raccoons and otters, the remaining snails were waging an epic battle for survival. Crawford said, "They were out of food. They were climbing on each other to lay eggs on each other's shells. They were floating on the water. The decomposition, it was intense. Rotting snails bubbling in the water."

Week by week, Crawford monitored the snail die-off. He told me, "I would go out and walk around and dig in the muck. There were places where it was still squishy muck that you would sink into, and you could feel around in that and find the snails. At first, with every step you stepped on one, and you heard a crunch. And after a while, what was crunching was empty shells. I did that until I stopped finding any live snails."

Finally, Crawford was confident the draw-down had decimated the snails. Now, did he dare to bring the water back?

After about four weeks with almost no water in the cell, Crawford started to reflood it. He did it very slowly and shallowly, so that any snails would be exposed to predators. Bringing water back in gradually also provided time for the populations of aquatic predators like crayfish to rebound.

His cautious approach to reflooding the cell paid off. Now, did he dare to replant it? Once the water reached the right level, he began planting. He said, "It took years to grow the veg back in that cell." Fortunately, the snail population remained modest after that. "There's been a couple of times when you look in and you see a lot of pink and you go check it out, but it was never as thick," he said. "The snails and the predators seem to be self-regulating. Now we have a better mix of plants. Performance is a little better, in terms of phosphorus removal, but it took a while."

Crawford told me that he only had to quell one other snail rebellion. It was a new STA called C-44 in 2021. The STA had been constructed on former pastureland, and they had brought water in to drown out the residual grasses.

I recognized what seemed like another perfect storm: lots of vegetation, no aquatic predators.

Crawford saw that I was catching on, and nodded his head: "Yep. The snails just boomed."

The invasive snails did not mind at all that it was pasture grass. As Crawford put it, "They really don't care what the plant is as long as it's underwater. You put that green stuff underwater and they're going to eat it. They'll eat the green parts of Brazilian pepper and leave the stems floating."

It had been winter, not prime egg-laying season for snails, but they didn't care. They laid eggs everywhere.

I said, "Flooding a former pastureland, I'm guessing that you'd start off with zero aquatic predators."

He said, "I mean, you'll get mosquito fish as soon as you bring the water in, and within a few weeks you'll get some little blue gills. Things just come in with the water. But the green sunfish and the things that are the real snail predators like redear sunfish, they're not there yet. And maybe the birds have decided not to hunt there yet. And the things that would be predators of the baby snails were not there yet, like insect larvae, crayfish, and leeches, and the different fish that would eat up a three- to four-millimeter snail. In a brand-new wetland, something that we just built, that used to be a pasture, those new aquatic predators hadn't had time to proliferate yet."

A light went on for me. I said, "I have heard the invasive snails described as a 'disturbance species,' that they like wetlands that have been 'disturbed.' But in ecology, disturbance can be anything that knocks an ecosystem out of balance, which could be a drought, a flood, or any number of things. I could not guess what kind of disturbance was key for the invasive snail."

Crawford nodded. "Perhaps part of the reason they have a reputation as a disturbance species is just that they like a wetland where their predators are not well established yet," he said. "It's a slower process to establish big stable predator communities than it is to establish the herbivores. Especially if you're going from zero to 'Here's a bunch of plants.'" He had witnessed two very different examples of disturbances that could lead to a boom. In STA-1E in 2013, it was murky water from construction combined with hydrilla. In 2021 in the C-44 STA, it was flooding of dry land. I wondered what other types of disturbances could suppress aquatic predators and provide plenty of food for the snails.

The second time around, Crawford saw the signs early and immediately dried the cell down. Once again, the predators had a field day. And once again, snail kites made a lot of nests. The C-44 STA was the second-most productive nesting site in 2021, with seventy-two fledglings.

Crawford said the STAs are now "self-regulating," with aquatic predators keeping the snail population in check.

We had been talking for a couple of hours when we reached the outflow end of the cell. I asked him if he ever saw native snails. He said he does; indeed, the balance between the native and invasive snails changes as the water quality changes from the inflow to the outflow. Invasive snails dominate at the inflow end of his cells, where the water is dirtiest and cattail flourishes. Downstream, as submerged vegetation took up phosphorus and locked it into the soil, the invasive snails would thin out. At the outflow end, he told me, the water was so clean that periphyton was abundant. That's where he saw more native snails than invasive ones.

I couldn't help interjecting, "Really? Calcareous periphyton?" It seemed to me that this gold standard of clean water would be hard to achieve with farm runoff, even after flowing through acres of submerged vegetation.

"Oh, yeah," he said, proudly. He stopped the truck to point out his window at the marsh below. "That's all periphyton right there. Do you see it?"

I could make out something buff colored in the water.

"Can we go in?" I asked.

Crawford grinned and said, "Sure!" He cut the truck's engine, and we both started creeping down the levee. On such a hot day, the cool water soaking into my sneakers felt great. Soon we were both knee-deep in the perfectly transparent water, our feet sinking into soft muck. Every third step or so, we'd encounter a solution hole, and we would sink thigh-deep into muck. The first time it happened I had to tug my leg out and then go back in for my sneaker. I wobbled as I balanced on one foot to tighten the laces and double-knot one shoe at a time.

When one of Crawford's Crocs got stuck in the muck, he retrieved it and took the other one off. He left them floating on the water. As we talked, I watched them float inch by inch downstream, an amusing reminder that the water is flowing, albeit very slowly.

The periphyton that I could see between the sparse, skinny blades of spikerush was unlike anything I had seen before. It was a thin coating on what I thought was submerged vegetation, grayish green and branched, undulating under the water. Crawford grabbed some of it and showed it to me in his hand, dripping with water. He said it is not a plant, but a green macroalga called muskgrass. Crawford said he can always tell itis around because it has a faint musky odor. I grabbed my own sample of it and breathed in its earthy smell.

Crawford pointed out that the periphyton film on the delicate muskgrass gave it a slight crunchiness. I rolled some between my fingers and felt a grittiness.

"That's the calcium carbonate in the periphyton," Crawford said.

I was fascinated to see a stormwater treatment wetland up close, and to hold in my hand the evidence of how powerful it was at removing phosphorus from the water before it enters the Everglades. The STAs are one of the biggest conservation successes that Florida can claim. I realized that Central Florida's lakes and canals are like the inflow end of the STAs: polluted water, invasive plants, and happy invasive snails. The Everglades are more like the bottom end of the STAs: clean water, native plants, few invasive snails, and good habitat for native snails.

"Now I get it. I have been trying to figure out how the invasive snail has taken over the lakes in Central Florida but never really got established in the Everglades National Park. In the Everglades they're mainly in the canals."

Crawford nodded, saying, "It would make sense."

I mused, "It's almost like, if we ever really wanted to get rid of the invasive snails, all we would have to do is clean up the lakes. Clean water is like invasive snail repellent."

"I mean," Crawford clarified, "I do see the invasive snails down here." He motioned to the muskgrass meadow we found ourselves in. "Just not nearly as many as in the upstream parts. And a lot more native snails."

FIGURE 13.1. Eric Crawford, shoe in hand, as he wades in the downstream end of the STA cell that had the apple snail explosion ten years prior in 2013. In the foreground, signs of high water quality: sparse stems of spikerush, and periphyton delicately coating the submerged muskgrass. Photo by Hilary Flower.

The slow-moving water had not taken Crawford's floating Crocs too far. He waded over to them, and I followed him out of the unexpected periphyton marsh at the bottom of the STA.

Now I was starting to understand why snail kite supersites caused by the invasive snail never seemed to last more than a few years: The predator populations caught up to them, a process which may take a few years in a natural wetland.

Suddenly Paynes Prairie made sense. I recalled that it all started with the whole landscape flooding in 2016. The newly flooded areas would have few aquatic predators, giving snail hatchlings a big survival boost. No wonder the snail kites came, and even made it their nesting supersite in 2019. Over the years, the aquatic predator populations caught up, and also water levels returned to normal. The boom started to turn into a bust. The snails may still be abundant, but not enough to support large-scale snail kite nesting.

The native snail did not have a reputation for exploding like the invasive snail. And yet back when snail kites were relying on the native snail, they found what they needed in places like WCAs on just a normal year. It was another important difference between the snails.

The next question I wanted to answer was, What is the invasive snail's origin story in Florida? It turns out that it started with the biggest boom of all. And snail kite conservation would never be the same.

# THE BIG BANG

# 14

"The whole edge of Lake Toho was just wrapped in pink. Snail eggs everywhere." Dr. Wiley Kitchens was recalling one of the most surprising sights he witnessed in his many years as the head of the UF Snail Kite Monitoring Program. Lake Toho is about eighty miles north of Lake Okeechobee, in the Kissimmee River Chain of Lakes Area in Central Florida (see figure 1.2).

Kitchens said, "You know that artist Christo, with the islands?" A quick Google search produced striking aerial photos of several islands near Miami wrapped in broad swaths of bright-pink fabric, a 1983 art installation called *Surrounded Islands* by Christo.

Kitchens said, "Well, that was what it was like."

Retired now, Kitchens spoke with me via video from his home in Jacksonville, Florida. He wore wire-rimmed glasses and had gray hair and a mostly white beard.

The Christo effect in Lake Toho marked the very public debut of the invasive snails in the fall of 2004. It wasn't their arrival per se. Apple snail expert Dr. Phil Darby's field tech, Dave Mellow, first detected them in very small numbers in a northeastern arm of Lake Toho called Goblet's Cove in 2001, 2002, and 2003. Like Florida's two other non-native apple snail species, this non-native species was keeping a low profile.

I tried to imagine what could have led the small number of non-native snails in Goblet's Cove to have a population explosion. Based on my education from Crawford at the Stormwater Treatment Area (STA), I guessed that something had kept predators away from the snails. But I couldn't imagine what could do that in Florida's seventh-largest lake.

Kitchens told me he knew exactly what the disturbance was: an $8 million habitat restoration project. They dried the edges of the lake, exposing most of the aquatic plants to the air. I didn't realize that you *could* turn a knob and partially drain a lake, but in Florida, many of our water bodies are heavily managed with water control structures.

After they lowered the water level in the winter of early 2004, bulldozers scraped seventeen thousand acres down to the sand layer, piling the muck into large mounds that became new islands. The scraping targeted a wide band of shoreline that would normally be one to three feet under the lake's surface. The water had become covered by thick mats of a native plant called pickerelweed, which can grow too thick for wading birds and other wildlife.

The goal was to make space in that band for well-spaced native emergent vegetation like spikerush and fragrant water lily, improving habitat for bass and other wildlife. That's what it had been like when Steve Beissinger found snail kites nesting there in the drought of 1982. The muck-scraping was intended to make a sandy-bottom shoreline for better fishing.

So, Event #1 was planned: a complete reset of the lake edge, removing water, vegetation, and muck.

Event #2 was entirely unexpected: Three hurricanes (Charley, Frances, and Jean) unleashed a torrent of rain, rapidly filling the lake. Powerful winds caused massive waves, churning up the lake and further disrupting its shoreline and vegetation.

According to Kitchens, these two events set the stage for the snail invasion. "The lake rebound was incredibly fast," he said. "And look who shows up on the block: the invasive snail."

"We got some very unexpected results," he said. "Who would have thought there would be that explosion of those invasive snails? No one I know. But it happened. And then, from there, those snails have made their way all the way across the state."

I asked, "Are you saying that if it weren't for these events, this snail would have stayed quietly in small numbers in Goblet's Cove, and we may never have heard of it?"

Kitchens nodded enthusiastically: "They were always there in the background, at some level. But the fact is, we took all the controls off them. The fish love eating the young snails, so that had always kept the snail numbers down. When they dried down the lake, you had fish in a small pool in the middle of the lake, and then the water comes back to the entirety of the lake, so you have a diluted fish population. We released the controls on those snails, and they just went to town."

Kitchens went on, "Uncertainty is such a big part of all these management activities, especially at the whole lake or system level. This isn't Lake Toho back in 1945, and it never will be again."

Kitchens explained that when an ecosystem is reset, this fits into an ecological concept called panarchy. When he was at the University of Florida, two ecologists, Dr. Lance Gunderson and Dr. C. S. Holling, developed the theory. It's complex, but for our purposes, it's about how systems periodically collapse and renew, and the unpredictability of change. In an email, Dr. Gunderson told me that they chose the term "panarchy" because "we were thinking about the mythical role of tricksters such as the Coyote, Kokopelli and Pan, and their surprising and destabilizing roles in lore." Realizing that snail kites have been dancing with the trickster helped explain why they were always surprising us.

In a panarchy cycle, you start with a mature ecosystem like Lake Toho in 2003: pickerelweed mats settled into the edges of the lake. Then an environmental disturbance comes along and clears the decks. In the case of Lake Toho, it was the partial drainage of the lake, the scraping of its edge, and rapid refilling: blank slate.

The next phase is reorganization. Pioneer species come in that take advantage of the emptiness. They tend to be able to grow rapidly without much help from the environment. This rebirth process is sensitive to slight differences from the last time the system had a clean slate, which can send it into a whole new direction. This is where the trickster can come in.

In the case of Lake Toho, the jumbo-sized invasive snails moved into the open edges. It seems plausible that the aquatic predators were still pretty sparse. And the vegetation dial was turned up. There was still cattail and bulrush in the three-feet-deep water, and the non-native snails crawled up and spread out when the newly opened areas were reflooded. And before long, the scraped areas bore a thick green carpet of submerged vegetation, especially tape grass.

No wonder the invasive snail populations exploded!

The panarchy model predicts that eventually, with time and growth, the new system starts to stabilize out a bit. This part of the cycle predicts the bust part of the invasive snail's boom-and-bust cycle: As aquatic predator populations rebounded, the snail populations came down from their extreme numbers. At this stage, we arrive at a maturing system once again, but it may be unrecognizable from the last one. In Lake Toho invasive snails were here to stay, and from there they spread to lakes and waterways across the state and beyond.

The concept of an ecosystem becoming fundamentally different from what it was before is often referred to as a "novel ecosystem." Even before the restoration project, Lake Toho had changed from its natural state due to levees and water control structures as well as invasive plants. The sudden large-scale proliferation of the non-native snails was a game-changer for the snail kite.

Hydrilla was about to take that novelty to a whole new level. To understand how, it is necessary to understand some key aspects of hydrilla. Hydrilla is originally from Asia, and in the middle of the twentieth century, Florida was its gateway into North America. People imported it as a fast-growing ornamental for the aquarium trade, not unlike the invasive snails themselves.

Hydrilla spread through the lakes in Central Florida in the 1990s, covering 75 percent of Lake Toho by 1994. From an ecological standpoint, hydrilla has its pluses (cleans the water, decreases mosquitoes, good habitat for some wildlife like ducks and fish) and minuses (shades out native vegetation like eel grass and spikerush; can lead to low-oxygen water; makes lake edges useless for wading birds).

Hydrilla has become the dominant aquatic weed in the state. It is also the fastest-growing. Hydrilla grows exponentially by branching and lengthening. People often say that hydrilla can grow an inch a day. In 2012 the Florida Fish and Wildlife Conservation Commission (FWC) decided to see if it was true. They started with a five-inch shoot of hydrilla and measured its daily growth for five weeks, comparing it to a five-inch shoot of pondweed. The hydrilla started by growing three inches a day in the first week. It was branching and growing in three dimensions, like an umbrella opening. By week five it had branched 157 times and was growing sixteen feet per day, and had grown a total of 266 feet. An inch per day turned out to be a huge underestimation: It was more like 7 feet per day. At the end of five weeks, the pondweed had grown forty-one feet, just over one foot per day, which would have been impressive if wasn't being compared with hydrilla.

Hydrilla is named for the Hydra of Greek mythology, an aquatic serpent with nine heads; when Hercules cut off one head, two more appeared in its place. This is fitting: Broken fragments of hydrilla can go on to form free-floating mats. In this way, the thick, free-floating hydrilla breaks the fundamental rules of submerged vegetation on two counts; it is supposed to be rooted, and to stay below the surface.

Lake Toho is deeper than many Florida lakes, which is why it still retained enough water for the snail kites to nest there during the drought of 1981–82. At that time, snail kites foraged for native apple snails among the sparse emergent vegetation at the shallow edges.

But hydrilla put out the welcome mat for snail kites in the deeper part of the lake. Hydrilla can grow in water up to thirty feet deep, because it has a high tolerance for low-light conditions. As it grows upward toward the sunlight, it fills the water column, forming a visible mat at the surface, not to mention the free-floating mats. If hydrilla is allowed free rein in a nutrient-enriched lake like Lake Toho, it can exceed ten tons per acre.

In the years following the big 2004 restoration at Lake Toho, hydrilla spread across even the deep parts of the lake. The hydrilla mats fed the snails, and like in the STA, they protected newly hatched snails from fish and other aquatic predators. Hydrilla mats also made it easy for snails to eat with their snorkels close to the surface, which in turn made it easy for snail kites to spot them. Snail kites flocked to the deep part of the lake. Kitchens said that snail kites could stand on these mats to extract the snails from their shells. This was a big boon to their foraging, because otherwise they would have had to fly a fair distance to find a post, shrub, or tree for this purpose.

Shorebirds like least sandpipers began to use the hydrilla as a mudflat. This reminded me of Eric Crawford describing a great blue heron walking on the muck-covered hydrilla in the STA. It's interesting to think that, aside from ducks (which love to eat hydrilla), the wildlife most benefiting were shorebirds and snail kites, neither of whom would have previously been found in deep water. Talk about a novel ecosystem. Hydrilla turned the lake functioning inside out. For the first and only time in their history in Florida, snail kites were deep-water birds. Once again, the snail kites' behavioral adaptations were rapid, unpredictable, and extremely important for their survival.

For Kitchens, the key point about panarchy is the uncertainty, the element of surprise. What Dr. Gunderson called the trickster quality.

Kitchens said, "There's all kinds of uncertainty in that, and that uncertainty provides a screen for selecting out who's going to make it at this next point over time. That's the simplistic version. That's the Wiley Kitchens part of it."

I realized that Jean Takekawa's 372-kite roost in 1985 was another novel ecosystem. The mining operation stripped away the Everglades marsh that had been there and created deep pits and islands. Emergent vegetation, wading birds and snails were screened out; Brazilian pepper and gators were screened in. Fortunately, snail kite roosting was also screened in, along with nesting for wading birds and at least one snail kite.

I also thought of the farmer's ditches where Takekawa found snail kites foraging on the other exotic apple snail in the drought of 1982. That was a novel ecosystem on an even smaller scale of both space and time.

And Lake Toho in 2004 showed that even a multimillion-dollar restoration project can produce a shocking outcome. The invasive snails were not the only surprise. By the end of 2004, muck again covered the sand, and the native emergent vegetation did not become established at the lake edges, so the restoration project did not meet its intended goals. We have much less control than we think we do when we act upon ecosystems.

"So these concepts of panarchy and novel ecosystems are interesting. What are the implications for the snail kites now?" I asked Kitchens.

"To remember that we're dealing with uncertainty," he replied. "We must do our best to come up with our best management plans, but we must remember that they are fraught with error. If we are planning to tweak the system, there can be a system-wide response that can knock all of this out. We need to keep our eye on the snail kites and remember that they are in a precarious state. We have to monitor them. How can you count on an invasive exotic to come in and save an endangered species from extinction? To put all our eggs in that basket? It's risky."

"So, panarchy is a reminder that things can change rapidly," I said. "That we can't just look at the kites and think, 'And then they lived happily ever after.'"

"Exactly, exactly," Kitchens said. "And part of that, you know, deals with scale, too. How much of the system is being altered? Are you intending to tweak it and not expect a response in the kites? Rob Fletcher, I think, has made this patently clear to the US Fish and Wildlife Service. He helped write the recovery plan that said, 'No. The kites are not out of trouble yet.' It's precarious, and we can see that, as different systems tweak on and off. There are going to be responses. And if we aren't careful, we're going to get a system-wide response, like a big drought, or whatever, and it can knock all this back."

Kitchens shook his head. "When you reset a system, wipe the slate clean, you never know what you're going to get; that's where panarchy is a bit more than just 'disturbance.' We need to be cognizant that the system is never static. Instead, disturbances or tweaks can trigger panarchy cycles, transforming the ecosystem. A novel ecosystem requires adaptive responses from us, recognizing that it does not play by the same rules as the prior ecosystem."

Kitchens's tone was urgent. His point seemed pertinent to all the snail kites' recent nesting supersites: Large-scale disturbances had triggered a surge in invasive snail numbers. The snail kite was not only dancing with the trickster panarchy; it was relying on it for survival.

Learning the shocking nature of the invasive snail's entry into the snail kite's story changed my understanding of the raptor's plight in several ways. First, I couldn't get over the fact that it was a fluke. The invasive snail's spread across the state had been so decisive, I had assumed it was inevitable. That's what invasive species do; they invade. Now I was learning that sometimes humans unwittingly play a part in determining which non-native species in a region end up becoming invasive.

Second, I had become aware that a new round of chaos and chance could come along and change the rules again. The same dynamic that brought this stroke of luck to the snail kites could reverse their fate just as abruptly. Something in our human nature makes us bad at predicting the future. We have a strong bias toward assuming that our current state was inevitable, and that it will continue in the future. And when we make a change to an ecosystem, we think we know what we're doing.

Although ultimately the explosion of the invasive snail in Florida became a godsend for snail kites, that part wasn't inevitable either. In fact, in the early days, back in the fall of 2004 on Lake Toho, the invasive snail looked like the biggest threat the birds had ever faced. To fully appreciate that challenge, and to see how they overcame it, we have to rewind to those first strange weeks, and watch the birds adjust in real time.

# FALLING SNAILS 15

The story unfolded in dramatic headlines. First came alarm: "Invasive Snail May Damage Diet of Rare Everglades Bird."

Then, just a few years later, came amazement: "Bird Evolves Virtually Overnight to Keep up with Invasive Prey." That particular article opened: "The federally endangered bird, the snail kite, was faced with an interesting dilemma: The exotic apple snail was good to eat, but about two to five times bigger than the native snail that the bird usually consumed. What's a hungry bird to do? Evolve—quickly."

The snail kites had grown larger bills. Had evolution *driven* them to rapidly develop larger bills so they could handle the larger snail? If so, how? And what did it mean? The possibility of seemingly miraculous adaptation had been the thread of hope that I set off to follow, with the snail kites as my guide. I knew I needed to get to the bottom of it. I wanted to talk to people who were there at the very beginning, back in October 2004.

Dave Mellow may have been the first to watch the snail kites encounter the larger snails. He arrived at Lake Toho, about thirty miles south of Orlando, in October 2004 to do snail surveys with students for Dr. Phil Darby. He called Darby that night and said, "The snail kites were catching these ginormous snails and dropping them!"

The snails were indeed ginormous, and the kites were dropping them—a lot. Mellow watched snail kites flying lower than usual when they carried the snails, as if burdened by the heavy cargo. It hurt him to see them drop the snail before reaching their perch. He told me when we spoke by phone that he couldn't remember ever before seeing a kite drop a snail, although he did know it could happen. But that day it *kept* happening, with big splashes that made it clear that the falling shells were not empty.

And the ginormous snails were everywhere. Mellow said he had never seen so many snails in one place. He and the students put down snail traps that were about one meter on each side, and then rummaged through the vegetation and muck to count the snails. They were getting about five adult-sized snails in each trap. They were so ubiquitous that when he walked a short distance to throw the trap again, he was crunching snails below his feet the whole way.

When Darby heard about this, he knew it was important. He didn't have funding to study this new development, but he outlined a four-day field study for Mellow to carry out, a simple comparison study. For the first two days, Mellow observed ten snail kites foraging at Lake Toho, where only jumbo invasive snails were available, no natives. The snail kites dropped eleven of the twenty-five snails they caught. And juvenile kites were the most likely to drop them. Then Mellow spent two days watching kites foraging in Lake Kissimmee, where the snails were all native. They didn't drop a single one.

I was eager to discuss this turning point in snail kite history with Dr. Phil Darby himself, recognized as the foremost authority on apple snails in Florida. He spoke with me by video from his office at the University of West Florida, in Pensacola, where he teaches ecology courses and runs a research program. With his scruffy goatee and baseball cap, he looked ready to head out to the nearest wetland. Darby told me that the field data made him think there might be a maximum size of snail that kites can handle well. The typical adult native apple snail is less than 2 inches (golf ball–sized), although they can grow as large as 2.4 inches (pool ball–sized). But in Lake Toho in October 2004, the *smallest* snail Mellow found was 3 inches across. The average size was 3.2 inches (almost softball-sized).

Snail kite talons evolved to close around golf balls, not softballs. And Mellow was right, weight was another issue. The invasive snails weighed five times as much as the native snails, and a third to half as much as the kites themselves! No wonder the snail kites were flying lower to the ground and dropping snails. The kites were also taking much longer to get the meat out, and sometimes they gave up. Their bills were curved exactly right to reach and snip the columellar muscle of the little snails (shown in figure 3.3). That muscle was thicker and farther back in the big invasive snails. The kites' fine-tuned bill no longer matched their prey.

It was strange that all the invasive snails along Lake Toho were jumbo-sized. In his paper, Darby surmised that the snails that had crawled up from the deep were the fully mature adults. If all the age groups had been there, the snail kites could have simply have selected for the younger, smaller ones.

Darby worried that while the abundance of snails might attract the kites to nest at Lake Toho, their young might not survive. In his paper on the four-day field study, Darby wrote that a detailed and thorough study would be necessary to calculate the trade-offs: Was the extra meat worth the extra energy expended to get it?

Fortunately, Dr. Wiley Kitchens, the director of the UF Snail Kite Monitoring Program, already had field techs and graduate students collecting data that could address that question. He had his team conduct meticulous time-activity budgets comparing kites foraging for the invasive snail in Lake Toho vs. the native snail in Lake Kissimmee and the Water Conservation Areas (WCAs).

Kitchens's field techs, including Brian Reichert (now Dr. Reichert) and crew lead Sara Stocco, put in some long days in 2006. They would take an airboat out into a remote area, pick the first juvenile snail kite they saw, and follow it all day. If it flew off, they would trail behind it in their airboat. At least one of them had to always have their eyes on it or their day's work would be wasted. At the end of the day, they returned to each foraging location to get a vegetation sample. Reichert, who went on to write his PhD dissertation on snail kites under Kitchens, told me that on the airboat he was typically up front, holding binoculars to his eyes with one hand and a stopwatch in the other. Behind him was Sara Stocco, with binoculars and a portable telescope with a stand, known as a spotting scope. Reichert and Stocco were logging every foraging behavior and marking down the time, much like Steve Beissinger had done many years ago. For this study, they were focusing on juvenile snail kites learning to hunt.

As concerning as it was to watch the snail kites fumble the snails, Reichert said the kites were finding tricks. One was to forage in the hydrilla in the deep part of the lake. Reichert brightened as he told me, in a video interview, "Over time, they were learning. Fewer snails dropped. Time for extracting snail meat got shorter. Their foraging efficiency increased. It was a novel system. The habitat was entirely different: deep lake with floating hydrilla rather than shallow spikerush marsh. I think there was social behavior that they were picking up. Kites' social dynamics are fascinating. They're clearly learning and communicating with each other."

Reichert went on, "They got pretty creative in making sure they wouldn't drop it. They'd get on PVC pipes, or hold the snail in certain ways. To get at the meat they must pick underneath the operculum, and then there's the columellar muscle. With the bigger snails, they couldn't get in there and pick it. But then they started to be able to get farther in there. I think the juveniles just didn't have enough experience with that initially, but they got there."

From 2003 to 2007, Wiley Kitchens's team amassed 935 observation hours of foraging snail kites. His student Chris Cattau crunched the numbers. The basic observations were similar to Dr. Phil Darby's study. Snail kites were ten times more likely to drop invasive snails vs. native snails. And it took them four times longer to extract the meat. That's a lot of wasted energy. And time. They ate a third to a quarter fewer snails per unit of time.

The burning question was the calorie trade-off. The invasive snails had about three times the calories of the native snails due to the size difference. And snail kites spent less time in flight when foraging for the invasive snails, which is one of the biggest energy costs to foraging. When Cattau calculated the gain of calories minus the estimated caloric cost of foraging, the results were mixed. The adult kites broke even, but the juveniles were in a deficit. In other words, juveniles expended more energy on the invasive snails than they got from the meat.

These results seemed to confirm what researchers feared: The snail kites appeared to be among the 42 percent of endangered species being harmed by an invasive species. And snail kite numbers were already low before this new challenge. Some kite advocates called for efforts to stop the spread of the invasive snails. But then, the handling problems went away. Because the size problem went away.

By 2007, snail kites moved into the deeper part of the lake, targeting invasive snails living amid hydrilla rafts. And for whatever reason, the snails were smaller there, Kitchens told me, based on the work of his student Lara Drizd. In places where a range of ages and sizes were available, the snail kites homed in on the same size class as the native snails, golf balls rather than softballs.

It turned out to be a good thing for the snail kites that no action was taken to suppress the invasive snails. Drought had brought the snail kites' population below seven hundred individuals in 2008. But the spread of the invasive snails helped their populations recover. By 2009, the invasive snail had colonized Lake Okeechobee and the other major lakes in Central Florida. By 2011, it was clear: The snail kites were staging a major comeback.

This is where Dr. Caroline Poli comes back into our story. She crunched data on more than 2,588 snail kites as part of UF's monitoring program, from 2002 to 2019. For young snail kites raised on invasive snails, it was like they were "born with a silver spoon in their mouth," as the saying goes. In fact, that's how she titled her paper: "An invasive prey provides long-lasting silver spoon effects for an endangered predator." Snail kites raised on the invasive snails weighed more at fledging than those raised on native snails. And their weight was less dependent on water levels being exactly right. After all, the invasive snails can thrive in a broader range of water levels. The invasive snails were freeing snail kites from their dependence on healthy ecosystems.

Most amazing of all: Poli found that the benefits lasted a lifetime. The kites reared on the invasive snails had higher survival even at ten years old, which is important for a raptor that can live thirteen years or more. And the survival difference wasn't subtle. The *top 3 percent* of snail kite individuals raised on native snails had the same survival rate as the *lowest 5 percent* of birds raised on the invasive snails. That's a huge boost in survivorship!

During nesting surveys, field techs would band older nestlings, weigh them, and measure their bills, leg bones, and other body parts—much as Steve Beissinger had done in the 1980s. When Cattau compared these measurements for birds raised on the invasive snails, he found bigger bills, longer leg bones, and heavier body weight.

This gets back to the question, What caused the bigger bills? Could it have been rapid evolution? Natural selection can happen within a generation if there's a mass die-off, leaving only those with key survival features. But, although the snail kites had struggled at first, the invasive snails had not sparked a mass die-off of snail kites. Quite the opposite. The invasive snails arrived when snail kites were in a nosedive and reversed their decline. Rather than eliminating birds, the invasives had greatly boosted their survival, as Poli's research showed.

The alternative explanation was simple: Better-fed birds are bigger. The nestlings' bills, along with their other body parts, could simply have grown larger when they were better fed. This reflects phenotypic plasticity—the ability of an organism's physical traits to change based on environment, without any genetic change. Children's height varies with nutrition, and sagittaria leaves grow larger or smaller depending on the nutrient content of their water.

Cattau coauthored a study that aimed to answer the question, Was it rapid evolution, or phenotypic plasticity? The study, which came out in 2017, concluded that the increase in bill size was "driven primarily by phenotypic plasticity rather than microevolutionary change."

When I had the opportunity to sit down with Dr. Rob Fletcher in a Gainesville café, I asked him to clarify the conflicting versions of how the snail kites' bills got bigger. When Dr. Wiley Kitchens retired in 2014, he hand-picked Dr. Rob Fletcher as his successor, and was "incredibly excited that he accepted." I explained to Fletcher that I was confused because the media touted it as rapid evolution, saying that the snail kites evolved bigger bills *in order to* eat the bigger snail.

"Basically," he said, "I wrote that paper. So, there's a backstory." He told me that, back in 2011, when the population of the snail kites was increasing, he started to think maybe it was because their bills were getting bigger. He said it took a while for him to talk his student, Chris Cattau, into combing through eleven years of their snail kite caliper measurements. But when he finally did, Cattau was excited to show Fletcher a graph making it clear that yes, their bills had gotten larger. Cattau did not have a background in evolutionary biology, so Fletcher had taken the lead in writing the paper.

"So, the backstory on that paper," Fletcher continued, "beyond working with Chris on it, is that we initially submitted that paper and said, 'Oh, there's evolutionary change going here. We've got all the parts.'"

He went on, "And, the reviewers came back and said, 'This is really interesting, but I don't believe it.' And they were very much . . ." He started again, with a slight wave of his paper coffee cup, "Because I'm not really an evolutionary biologist either. And the reviewers were. And they basically said, you need to look at your information in a different way. And when we did that . . . I think a simple way to think about it is that we were seeing bill change happen at a rate that was faster than can be predicted based on all the bits of microevolution that we were able to document."

It sounded like they weren't able to provide direct evidence for microevolution, so phenotypic plasticity was the default interpretation. And that Fletcher was still a believer. He told me that now that they had several years more data, he planned to revisit the question of rapid evolution. I resolved to keep an eye out for future papers.

As I walked back to my car, I found myself reflecting on what I'd learned. The rapid evolution story had brought me here, but somewhere along the way, the real story had emerged. I crumpled my paper coffee cup, realizing that I no longer felt any attachment to the idea that the snail kites had rapidly evolved. It had sparked my initial curiosity, but the snail kites had led me to their true superpower: behavioral adaptability.

They had weathered fluctuating snail populations through ambisexual mate desertion and statewide dispersal. They could survive in roadside ditches, farm canals, abandoned mines, the deep waters of Lake Toho, hydrilla-raft perches—nesting year-round when snails allowed instead of following their traditional season. And they had almost entirely let go of their two namesakes: the Everglades and the native snail.

Their bumpy transition to the invasive snail offered a rare glimpse of what adaptation looks like in real time. Harnessing the power of invasiveness had become their most stunning and hopeful adaptation of all.

Now I understand the invasive snail, and how the snail kite adapted to it. The first two mysteries were solved. But as I learned more about that transition, I came to a troubling realization: the biggest danger for the snail kites was not environmental, it was us. By 2008, it would be clear just how much havoc we could wreak on nesting snail kites. In the early years, though, the field techs were the ones bearing witness—and facing mortal danger.

FIGURE 16.1. Brian Reichert stands waist-deep in the cattail marsh of Lake Toho in 2007, holding a nesting pole topped with a mirror to check for eggs or chicks in the snail kite nest visible above and behind him as part of his work for the Snail Kite Monitoring Program of the University of Florida. Photo by Kyle Pias.

# THE DANGERS OF LAKE TOHO

# 16

It was 2007, three years after the invasive snail had transformed life on Lake Toho. Brian Reichert, wearing cargo shorts and a full-brimmed hat, was wading in water up to his chest in Lake Toho. He was there to check snail kite nests. Reichert was a member of the University of Florida Snail Kite Monitoring team, still led at that time by its first director, Dr. Wiley Kitchens. Reichert held a pole with a mirror on the top for seeing down into nests. Were there eggs? Were there young? What stage of development were they at? Repeated surveys every two weeks made it possible to evaluate snail kite nesting success.

Reichert said, "I opened up the wall of cattail, and a ten- or twelve-foot alligator is sitting there, perched right up on top of me. And then it came down into the water by me."

"What did you do?" I asked, horrified.

Reichert laughed, shaking his head. "All I had was the nest pole," he said. He gestured as if rowing an oar. "I used it to divert the gator away from me." On our video call, Reichert appeared in business casual, safe in his corner office at the US Geological Survey in Colorado.

Instead of fleeing to the safety of the airboat, Reichert just kept going. It was his first nesting season. He would go on to get his PhD and then a postdoctoral fellowship under Kitchens. He said Wiley Kitchens was big and sometimes gruff, and he had no tolerance for incompetence. Reichert loved working with him; he thought of him as a father figure. When I relayed that to Kitchens, he said, "Well, now you're going to make me cry." When Kitchens can't sleep at night, instead of counting sheep, he sometimes lists all of his students and postdocs. He was proud of them all, and Reichert was one of his brightest.

Joisting with gators was not exactly what Reichert signed up for. The field crew lead, Sara Stocco, trained Reichert in in the marshes of Water

Conservation Area (WCA) 3A, which is just north of the Everglades National Park, and about fifty miles south of Lake Okeechobee. In the heart of that 786-square-mile wetland, you can spend all day airboating as far away from human civilization as possible in Florida. It was a rain-fed periphyton wonderland like Indian Prairie and Grassy Waters, wild and remote enough for the gators to be shy. Reichert said the gators he saw there were usually on the edge of saw grass patches, canals, and alligators' trails. Fieldwork in such places was pretty safe if you paid attention. Snail kites nested in small shrubs and willows, making it easy for them to forage in the surrounding marsh. That made it easy for the field techs to check on them while keeping an eye out for gators.

In his early days of training in 3A, Reichert got to witness more than one snail kite dropping straight down from their perch to grab a snail, repeatedly, three or four times in the same location. The native snails were so abundant that it was like shooting fish in a barrel.

Over the next couple of years, though, Reichert witnessed kite foraging in 3A go downhill as the native snails declined. He said, "We would be chasing a single kite for a long distance, on these long linear paths, as opposed to staying localized. We'd be driving and driving, for what felt like forever, as the kite was just scanning and scanning. The kites weren't able to see any snails, that was my assumption." My mind's eye flashed to the long and often fruitless snail kite patrols I had witnessed at Paynes Prairie.

Reichert was witnessing firsthand how the scarcity of snails in former strongholds like 3A drove the snail kites to abandon the Everglades.

And Lake Toho was their number-one refuge.

Unfortunately, the gators there were different. Thinking of them lurching out at Reichert, I said, "I can't believe you had to deal with that in your fieldwork."

Reichert shook his head, laughing. "Oh, yeah. Definitely. It happened all spring."

When it comes to gator safety, people often say, "If a gator chases you, run in a zigzag." The fact is that alligators can move so fast that your path is irrelevant. The only way to stay safe from gators is to prevent confrontations in the first place. This means staying out of all freshwater in the human-dominated landscape. Gators who are used to people are also used to being fed and harassed, so they are much more aggressive. Don't put a toe in canals, retention ponds, golf-course ponds, or lakes of any kind, and keep your small dogs and children away from the edges. Even in remote areas with shy gators, like where I take students, awareness of your surroundings is essential. I stick to water that is clear and unobstructed by vegetation, so I can see where my feet are going. Stay out of gator holes. Give gators a wide berth if you do see them. Stay away from nesting habitat (like dense cattail). Avoid baby gators; their mothers protect them for two years.

But to monitor snail kites on Lake Toho, though, Reichert had to ignore all these cautions. This Central Florida lake hums with Jet Skiers, anglers, and recreational boaters. It is rimmed by docks, marinas, residences, tackle shops, bars, and restaurants. Despite the law, some people feed alligators hot dogs and marshmallows, teaching them to link humans with food. Others provoke them—throwing objects, trying to catch them, hooking them on fishing lines, even playing with their babies. The lake itself is murky and deep, and alligators nest in the same stands of dense cattail that snail kites do. It's the perfect recipe for trouble.

Snail kites prefer to nest in willow or small trees out in the open, like they did in 3A and Moonshine Bay, but cattail is more common on Lake Toho, and snail kites make do. Dense cattail was not even the scariest nesting habitat. Sometimes kites would nest on a tree surrounded by a floating bog of hydrilla. Reichert would stop the airboat five hundred feet away, so as not to disturb the kites, and walk the rest of the way. On the hydrilla.

Reichert recounted, "You're walking on this floating mat of vegetation," he said, "and breaking through it constantly. And there's just growling everywhere."

The foraging surveys put Reichert in even more danger than the nesting surveys. Whenever a snail kite that they were tracking would pluck a snail out of the water, the field techs on the airboat would mark the spot with GPS. At the end of the field day, they would go to every spot and get a vegetation sample. Back in the shallow water of 3A, Sara Stocco had shown Reichert how to sample the plant by your feet. Easy. But in Lake Toho, you had to dive into murky water well over your head and get the vegetation as quick as you could. Even more disturbing, by the time they were doing their vegetation sampling, it was evening; gators are most active at dawn and dusk. Reichert told me, "There's alligator eyes everywhere. And we're like, 'Well, I guess we gotta get the data, right?' And so, you jump into the water and do whatever you need to do to get data." If it got too dark to see the vegetation at the lake bottom, they just had to feel around.

Reichert went on, "Back at the office, we were telling Kitchens that we were afraid. Like, 'I don't think these foraging surveys are safe.' And he was like, 'Why aren't they safe? I don't understand.' And when we explained, he was like, 'Oh, my God! I didn't know that you guys were going like fifteen feet deep in the middle of the lake at eight o'clock at night!'"

At the end of field days, Reichert and Stocco washed their feet off in the shallow water of the boat ramp, where anglers often clean their fish. It was as if a dinner bell had rung. He said, "The gators would just come out of nowhere and start coming for us."

Reichert's closest encounter was on an airboat in a canal. As he sat pretzel-legged on the bow of the boat, eating a peanut butter and jelly sandwich, a five- to six-foot gator launched itself halfway onto the boat. He said, "We pushed it off with a canoe paddle that we kept on the boat, in case we got stuck or had to paddle back to shore."

Reichert had a recurring nightmare his first summer that he would be diving down, and a gator would emerge from a ledge in the white limestone bedrock at the base of the lake and come after him. On canals he would often see alligators resting on karst shelves as the airboat went over them.

But over time, he got used to it. Having learned what it took Reichert and the other field techs to get snail kite nesting and foraging data, I started to read studies coming out of that lab with extra admiration and gratitude.

Yet the only thing more unsettling than jump-scares from gators in hydrilla and cattail was that these habitats could disappear overnight—even when the snail kites were actively using them.

# NUKE IT 17

Dr. Brian Reichert told me that while checking on snail kite nests, he would commonly see young birds, ten or eighteen days old, not big enough to fly. "And then," he said, "we'd come back a couple of weeks later for the next check, and the cattail would be dead, and the nest was falling over, with no sign of the young. Sometimes there were adults still squawking nearby."

There could be little doubt that Florida Department of Environmental Protection (often referred to as DEP) was behind these incidents. Reichert said, "I remember twice we were out there, and the DEP guys were spraying herbicide right by the nests!"

Cattail is a plant native to the Everglades. In moderation, cattail provides benefits. It filters nutrients out of water and stabilizes the soil. It serves as valuable nesting habitat for animals like common gallinules, grackles, rails, and bitterns, not to mention alligators. It can also provide habitat for frogs, lizards, worms, insects, and small fish. And as we've seen, kites can use it for nests where willows or other small trees are scarce, like in Lake Toho.

But "the dose makes the poison": A little cattail is beneficial; a lot becomes harmful. Where water is deep and nutrient enriched, cattail can become so dense that it outcompetes all other plants, resulting in a cattail monoculture. The lack of sunlight hitting the water prevents periphyton from growing—there goes the base of the food chain. Dissolved oxygen in the water tends to be low, so fish and bug densities are low, further limiting the prey base. Besides, few birds could squeeze through.

When snail kites do nest in cattail, they need it to be a small patch surrounded by open marsh for foraging.

Unfortunately, cattail has closed over thousands of acres of marsh in South Florida, obliterating the wildlife that relies on open marsh. As a result, cattail is one of the few native species that are treated like invasive species.

Sometimes helicopters dropped pellets of herbicide over many acres of wetlands at one time, sparing neither the nests, nor the field techs. Dr. Wiley Kitchens got a call from one of his field techs out in the cattail, saying, "I'm out here monitoring nests, and they are dropping pellets of herbicide all around me!" All those nests failed. To Kitchens, this was a clear violation of the Endangered Species Act (ESA).

When Kitchens complained, the DEP just said, "Prove it."

Kitchens's group reported them to the US Fish and Wildlife Service (USFWS, sometimes referred to as "the Service"), which enforces the ESA. The Service could sometimes be a go-between for the kite monitoring team and the DEP. That's what happened in this case.

Reichert told me, "After that, every time we did a nest survey, we would update our nest maps for the invasive plant crews to know where to avoid, and the Service would take that information and pass it to them." Designating a buffer zone around all snail kite nests was an important measure, but it wasn't always enough.

The drought in 2007 highlights the snail kites' increasing dependence on Lake Toho. The drought cut the snail kite population in half, as did the one in 2001, and the droughts that Steve Beissinger and Jean Takekawa witnessed in the 1980s. And much like those years, Lake Toho became an emergency wetland. With the inaugural explosion of the invasive snail there in 2004, Lake Toho had become an emergency wetland on steroids. Kites on Lake Toho not only survived during the deadly drought of 2007, but they also produced seventy-three of that year's eighty-one fledglings. It was a stunning bright point in a deadly year. It was as if Lake Toho was in another universe.

Kitchens knew that it wasn't sustainable for the snail kites to keep losing half of their population. He projected snail kites had an 80 percent probability of going extinct in the next three decades. His team had shown that adult snail kites are long-lived; in a typical year 80–90 percent survive. In fact, in 2000 Kitchens's crew spotted one of Steve Beissinger's 1979-banded snail kites, a female. She was twenty-one years old, and trying to nest. Snail kites can live upward of twenty-four years.

Kitchens argued that the factor most critical to snail kites' survival as a species was nesting success. He told me, "Our monitoring data showed reproduction was key. You wouldn't expect this from a long-lived bird, but it's true. They were losing their reproductive capacity. They were losing sites for nesting." Kites' foraging in hydrilla and nesting in cattail put them at odds with the DEP's agenda for Lake Toho. "Weed management," Kitchens said, "was the agency action that was most threatening to the kites' reproductive success."

This all came to a head in Lake Toho in 2008, when the state's biggest site for snail kite nesting collided with the state's biggest hydrilla problem.

Lake Toho hosted eighty-four kite nests that year, which comprised 61 percent of the kite nests statewide. If those nests failed, it could drastically steepen an already steep decline. To prevent this, the kite monitoring crew from UF had shared their nest locations and protective buffer maps with the DEP. Kitchens was alarmed when he learned that snail kite nests were in the zone of a huge hydrilla treatment coming up. Many of the nests were weeks away from fledging young.

Kitchens's crew could see the horror movie play out: the hydrilla dying all around the nests, the snail kite parents flying long distances to procure snails, while their young starved to death or were taken by predators. They had seen it before, just never on this scale.

But the DEP was up against a formidable hydrilla problem. By the spring of 2008, hydrilla mats were visible across Lake Toho's nineteen thousand acres. At public meetings, tempers flared. Hydrilla was entangling engine propellers in boats and clogging jet skis. Big sections of the lake became navigable only to airboats that could skim across the hydrilla surface. Lakefront residents complained that hydrilla made it impossible to get out onto the lake. Duck hunters had the highest tolerance for hydrilla, since ring-necked ducks and other desirable ducks love to eat hydrilla. But the duck hunters did need channels cut through it to get to the best duck hunting spots. Anglers, too, could benefit from moderate hydrilla because bass, bluegill, speckled perch, and baitfish thrive under its shade. But when allowed to grow out of control, hydrilla can block their boat access and even close over fishing holes. In turn, tackle shop owners, fishing guides, and marina owners can become collateral damage.

The Army Corps of Engineers was concerned that hydrilla could clog up the canal bringing water into the lake, leading to flooding. Flood control was their biggest mandate at Lake Toho.

Municipal lakes must balance many conflicting demands. The two agencies charged with solving this problem were the DEP and the South Florida Water Management District (SFWMD). They were used to these battles. Their mandate from the state was clear: Eradicate hydrilla. And that's what they planned to do.

The two main methods for controlling hydrilla are chemical and mechanical, each with trade-offs. Herbicides can kill it all, but the dead vegetation releases nutrients and causes lethally low dissolved oxygen levels as it decomposes in the water. By 2008 hydrilla had developed resistance to the most-used herbicide at that time, Fluridone, causing invasive plant managers to use up to ten times higher concentrations. Those high doses could wipe out the entire habitat. Worst of all, these chemicals could be unpredictable as well, sometimes drifting far from the intended kill zone.

Mechanical removal avoids all these problems. Unfortunately, it captures fish and other aquatic wildlife along with the hydrilla. And it's like mowing the lawn; the harvesters remove the top five to ten feet, leaving the base of the plant to regrow soon thereafter.

Other methods include introducing grass carp or other grazers. But grass carp are just as hungry for native plants as they are for hydrilla. It's unfortunate that the invasive snail, though numerous and voracious, could not make a dent in the hydrilla.

I'm guessing it was the budget that tilted the plan in favor of chemical treatment. Mechanical removal costs two to three times more. They had a multimillion-dollar budget, but that was only enough to eradicate hydrilla in a fraction of the state's affected wetlands on any given year. Meanwhile, hydrilla would continue to expand in the lakes that didn't get treated, making the DEP hit them harder when they did get to them. Every year they played whack-a-mole. So, they were highly motivated to maximize the areal coverage of their hydrilla treatments with the money they had.

In 2008 it was Lake Toho's turn. Helicopters would drop heavy doses of Fluridone on thousands of acres, including eighty-four active kite nests.

They were going to nuke it.

# 18 THE HYDRILLA WAR

The Hydrilla War of 2008, that's what I started to call it in my head as I pieced events together. The way it played out tells you everything you need to know about the agency-level barriers and opportunities when it comes to saving endangered species. And it provides a vivid and surprising answer to the question: How did the invasive snail change snail kite conservation?

The turning point in this battle was Nat Reed.

Nat Reed: People say his name like it's a full sentence. Nathaniel Reed, who died a legend in 2018, was an old school power broker and a Republican. His family owned the better part of Jupiter Island, a large barrier island north of West Palm Beach.

Nat Reed was "a very bullish person," as Dr. Paul Gray of Audubon Florida put it. Everyone spoke to me about Nat Reed with a kind of awe and humor, like, "Wow, that man."

And this force of nature was an environmental crusader. He cowrote the Endangered Species Act. He had a hand in stopping DDT use, passing both the Clean Water Act and the Marine Mammal Protection Act, establishing Big Cypress National Preserve, and expanding national parks.

And when Reed retired to Florida, he fell in love with the snail kites. Before long, he was an honorary member of Dr. Wiley Kitchens's snail kite monitoring crew, often going along for surveys and flying over in helicopters with Kitchens's graduate students.

As the hydrilla treatment in Lake Toho approached, Kitchens was tempted to bring Nat Reed into the fight. But he knew that all hell would break loose, and he would face heavy backlash for not going through the proper channels. But the proper channels were not working. As it happens, Reed called Kitchens to ask for a favor, and he decided to go for it. He told Reed about the treatment plan and the nests.

"Wiley," Reed said, "leave that to me."

Kitchens laughed as he recounted this to me. He said, "Nat lit the world on fire. The next morning, I had messages from Washington, DC, Atlanta, Tallahassee, and West Palm Beach."

Nat Reed detailed his own account of the events in a 2010 letter that Paul Gray shared with me. In Reed's telling, he first called up the heads of the state agencies, "to halt the planned assault on the hydrilla."

Next, Nat Reed called Paul Souza at the US Fish and Wildlife Service (USFWS). Souza was the person that had to approve of such plans. Reed wrote that Souza "responded immediately with telephone calls invoking the Endangered Species Act." This law (the ESA) bars government agencies from doing anything that could jeopardize the continued existence of a listed species.

The agencies were certainly listening now. They drafted a new plan. They would cut way back on their planned treatment zone, providing a wide margin around the kite nests. The DEP put the treatment back on the calendar, confident that all would be well.

Kitchens didn't share their confidence. And he wasn't the only kite advocate with reason to be wary of the plan. Dr. Paul Gray from Audubon Florida had recently witnessed how bad treatments like this could go. There were two herbicidal treatments at the lake nearest Gray's home, Lake Istokpoga. One targeted hydrilla in the middle of the lake. Herbicide application is not an exact science, especially with the tools available at that time. Temperature and currents can influence the potency and reach of chemicals. The kill zone on Lake Istokpoga had expanded to twice the intended size. Fortunately, there weren't any snail kite nests there at the time, but Gray was appalled by the ecological damage.

A second treatment at Lake Istokpoga, on another year, targeted the emergent marsh, all the way from where it was dry enough for trees to grow, to the deeper parts where only plants like bulrush could flourish. Gray told me, "They killed all of the plants from the tree line out to the bulrush line, turning the entire emergent marsh around the lake into open water."

Gray confronted the invasive plant manager, who asserted that it was his job to get rid of an invasive plant in the emergent marsh and that's what he did. "But now you don't have *anything,*" Gray responded. "I mean nothing can live there. There's no frogs, there's no snail kites, there's no snails. You killed everything just to get rid of this one plant. Your treatment was worse than the disease."

With that experience fresh in his mind, it is easy to see how Gray could be wary of even the revised hydrilla plan for Lake Toho.

On June 10, 2008, they treated three thousand acres of hydrilla on Lake Toho. Everyone waited to see the results.

The results made Nat Reed furious. He wrote, "Herbicides were used in 'distant' portions of the prime snail range, but it is clear that the herbicides migrated toward the prime feeding areas and the colony lost a number of young birds to starvation." Thirty-six of the eighty-four nests on Lake Toho failed that year.

Worse, it took years before the hydrilla rebounded enough to make Lake Toho as successful for snail kite nesting as it had been. Since Lake Toho had been the biggest nesting site for snail kites for the last few years, and since snail kites had lost half of their population twice in less than a decade, this was a tremendous blow to the fight to save snail kites from extinction. Snail kite advocates continued to fight, and some key things changed.

In July 2008, responsibility for invasive plant management passed from the DEP to the Florida Fish and Wildlife Conservation Commission (FWC). The FWC created a new position, the snail kite coordinator, and set up a formal protocol for better enforcing the buffers around snail kite nests. As before, Kitchens's crew would scout nest locations and record them with GPS. The snail kite coordinator would place two circles around each nest location on the map. The outer circle was a low-activity zone (with a radius from the nest of 500 meters, or 546 yards), and the inner circle was a no-activity zone (500 feet). Any government agency planning to do something within those zones would have to go through the snail kite coordinator, who in turn sought the guidance of the Service.

The first coordinator was Dr. Wiley Kitchens's former student and veteran nest checker Dr. Zach Welch, who had completed a PhD and postdoc under Kitchens.

In a video call, Welch told me, "When I was at FWC from 2009 to 2014 and the population was doing really, really poorly. We didn't want to do anything, I mean *anything*, that could be jeopardizing at all." He said that when the Service said something needed to happen for the snail kites, the

FWC policy was to follow it as exactly as possible. The FWC wanted "no daylight" between their actions and the Service recommendations.

"The majority of my time there," he continued, "I was coordinating hydrilla treatments and other habitat management treatments like cattail." He said it was a time of breaking down silos, where agency people with varying expertise would discuss actions that might affect the snail kites. He said, "We would go out on a boat together, and everybody would be there talking. What does this do for wading birds? What does this do for fish? What does this do for snail kites?" This new position was the most important and lasting outcome of the Hydrilla War. Welch in his day, and then his successor, Tyler Beck, advocated for the snail kites every day. When I was tempted to become cynical about agencies, I tried to remember the ongoing, quiet efforts of many people on the ground, doing what they could.

Still, there was one word that guaranteed conflict. Dr. Paul Gray of Audubon Florida explained, "Originally the legislature directed the invasive plant managers to 'eradicate' hydrilla; that's the language, 'eradicate.' And we were fighting with the invasive plant managers, saying, 'Guys, you're eradicating kites along with it!' And they were like, 'Doesn't matter, we were ordered to eradicate it, we have to do it. We're legally bound to do that.'"

Alas, very often laws are written by lawmakers without scientific training, so there's a mismatch between what is needed on the ground and what passes into law.

Gray wrote a compelling argument and shared it with all involved agencies. He pointed to research showing that marshes with hydrilla had two to six times as many snails compared to marshes without it. He showed that the 2008 herbicide treatment at Lake Toho damaged snail kite reproduction for years to come. In 2010 only eighteen of the seventy-nine nesting attempts at Lake Toho fledged young, a failure rate of 77 percent.

Gray said his paper boiled down to, "Would you please quit treating this invasive plant, because it has an invasive snail that can save my endangered species!" He laughed and added, "The press had a heyday with that, because that's a weird question."

Novel ecosystems call for novel conservation strategies. I'm sure that when Gray joined Audubon Florida, he never imagined spending so much time fighting to save *invasive* species.

Nat Reed's 2010 letter accused the government of failing to protect this endangered species. And he was prepared to escalate.

The law changed: "Eradicate" changed to "manage." Snail kite advocates won that round.

Gray told me that in 2011 something miraculous happened: "They got enough political pressure that they didn't spray Lake Toho that year. We got seventy-three babies out of that one site. That's amazing, especially if they survived their first year. It was a banner year. We only had seven hundred birds, and we got seventy-three babies."

# THE CATTAIL STUMPS AND THE BABY KITE

# 19

The hydrilla in the deep part of Lake Toho was just the first in a series of unexpected habitats to become snail kite nesting supersites in the age of the invasive snail. Oddly enough, one of the strangest is Lake Okeechobee. The only part of the snail kites' Critical Habitat that they still use (see figure 5.2), it has become an ecosystem in a cycle of collapse, and an accidental success for the snail kite. For some, it's a cause for celebration. For Dr. Paul Gray of Audubon Florida, it's a disaster. He often brought it up to me, saying, "It was a moonscape. They killed everything."

I realized that to understand the plight of the snail kites, I had to understand what was going on in Lake Okeechobee.

It started with a ten-thousand-acre rectangle of cattail monoculture in Moonshine Bay, part of Lake Okeechobee's western marsh (see figure 1.2). In 2015, the aquatic plant managers from the Florida Fish and Wildlife Conservation Commission (FWC) helicopter-dropped herbicide pellets all over that rectangle and then came back to burn off all the debris. The result: blackened stubs poking out of open water. A ten-thousand-acre zone of death.

In the aftermath was a new beginning. Nutrients from the ashes promoted new growth. Light could finally penetrate through the water again, which in turn allowed algae to grow on the dead stalks under the water. Submerged aquatic plants started to grow. Snails climbed up the dead cattail stalks to breathe and to lay eggs. Aquatic snail predators like fish would not have been numerous in the dense cattail, so for a while the open water was practically predator-free.

Abundant snail food, scarce aquatic predators: It was the perfect recipe for an invasive snail population boom, and they sure boomed. Snail kites took full advantage. On Lake Okeechobee, 2016 was a record year for snail kite nesting: 255 fledglings, more than ten times the number in the previous year.

I had to see it for myself.

Most of all, I wanted to see it through Tyler Beck's eyes. He had spent many years protecting snail kites' interests on the lake as the snail kite coordinator for the FWC. The treatment would have required his approval. I wanted to see what a snail kite nesting supersite looked like in the era of the invasive snails. When Beck offered to take me out on the lake, I leaped at the chance.

On a scorching summer morning, I found myself seated next to him on an airboat as it bumped and skimmed over the water. His sunglasses shone iridescent-blue below a dark-gray baseball cap. He could have been taken for just another angler on the lake until he stood up and the prominent "FWC" became apparent on his back. We started on the tree-lined rim canal that runs along the inside edge of the Herbert Hoover Dike. Then we took an offshoot to the east, Moore Haven canal, which started off narrow and soon opened out to a water world. The blue sky and clouds were reflected in the water as we passed patches of low vegetation to either side.

Beck directed the airboat straight into a wall of cattail. The cattail just lay down and popped back up behind us as we passed through. We found ourselves in a liquid meadow of white lilies with delicately notched lily pads. At the other end of that pond, we motored around a stand of willow to find ourselves in what looked like a grassy savannah stretching off to the horizon. As we skimmed over the grasses, I saw sparse sagittaria here, purple blooms of pickerelweed there. My eyes were filled with so many different marshes, each one beautiful in a new way.

And the kites! So many snail kites flew across our path or swooped around above the marshes that I stopped counting at twenty. It was like fireworks bursting overhead.

Beck stopped the airboat in a big patch of fragrant water lilies with a border of willows. He told me we were in the 2015 treatment area. The abundance of snail kites was a direct effect of the 2015 treatment, and the subsequent upkeep.

Beck explained that the original hope with the 2015 treatment was to return the whole area to the spikerush marsh of the past, but that mostly didn't happen. Still, a cattail monoculture had been replaced by a mosaic of different wetland habitat types. A greater diversity of wildlife was flourishing there. It was hard to believe that the lively patchwork of wetlands we'd been flying through had been barren just a decade ago. Beck told me they still had to come and do spot treatments to keep it open. The deep, nutrient-enriched water ensured that the cattail would always come back and killed off in an endless cycle.

A short distance away, he showed me a recent treatment area. There was no vegetation, just open water and straw-colored stems of dead cattail sticking out a few inches from the water, some pointing straight up, many leaning at odd angles. Beck slowed down and stopped the airboat.

A brown snail kite made an irritated call and flew over to a tuft of willow about fifty feet away. Beck stood up in the flat front part of the airboat. Peering through his binoculars, he said, "I'm a little suspicious of this female here. I was in here a couple of weeks ago. I saw some birds carrying some sticks." As he sat back down in his captain's seat, he said, "Now, when she pulled up here, she kind of cackled at us. It may just be like, 'Hey, that's my favorite foraging perch and you've pushed me off it.' Or it could be that she has more stuff going on."

I said, "Like nesting?"

"Could be," Beck said.

He told me there were kites nesting nearby. I was afraid to hope that I might get to see my first snail kite nest, especially since it was August. I searched the willow for signs, but still didn't entirely know what to look for.

The snail kite abruptly flew off her perch and snagged a snail from a nearby cattail stump. I realized that since the snails needed to crawl up the cattail to breathe, these stumps were giving them away to the snail kites.

Beck explained that the scene before us was due to this past winter's treatment of a dense stand of cattail here. The invasive plant managers had planned to follow up with a burn before snail kites nesting season got under way. But they couldn't: The snail kites had already moved in and started nests. He said, "This year, over half the nests in the state were here in Okeechobee, because not many other areas performed very well."

This was the first truly open water we'd come to after we left the canals. The rest had hosted vibrant vegetation. Gazing around at the buff-colored cattail stumps, I figured this was what Paul Gray meant by the moonscape.

As if reading my mind, Beck said, "I'm not sure Paul Gray has been out here in a long time, but he was out here shortly after we did the big treatment in 2015. It looked a lot like that." He motioned to the open water with sad stumps of cattail. "Except we were able to burn off the standing dead cattail over there, and we weren't able to here."

Beck went on, "Paul thinks we just obliterated this area. Well, we kind of did at first. It didn't look nice for the first year or two."

For his part, Beck saw the glass as half full. He said, "I'm pretty happy with this area." He went on, "We're kind of doing this game right now, where we're doing a treatment over here, and it gets good nesting for a couple of years, and then we do a treatment over here. . . . It's not the ideal management scenario, but we aren't doing these treatments simply for snail kites. We're trying to manage the habitat to benefit other species as well." Considering the low biodiversity in cattail monocultures, it seemed like anything would be better.

Beck continued, "Normally, natural fires and droughts would come through and wipe out vegetation, opening it up and allowing a new round of vegetation to grow in. But those things no longer happen on the lake anymore. So, we are using other methods to create the patches."

I had to think about that for a moment. Lightning from the summer thunderstorms used to ignite marsh vegetation, which was flammable above the water. Fire was good for the marsh, because it suppressed woody vegetation, released stored nutrients, and kept the marsh open. The water protected the roots from burning. But although dead cattail will burn well, live cattail does not, especially not when it's packed tightly. Deep water inhibits fire further. The FWC had taken on the role of disturbance creator, resetting the habitat back to open water repeatedly.

Beck said that when they modify the habitat in some way, the invasive snails go crazy, and then subside to a lower level over the course of a couple of years. I thought about how the native snail could not survive these conditions. I said, "It's a good thing the snail kites have moved over to the invasive snail."

Beck took this up. He said, "A lot of time the habitat treatments are making things really good for the invasive snail, and that's not really our goal. But the secondary benefit is that snail kites are basically persisting on an invasive snail right now."

He smiled and said, "I got a phone call one day from somebody from Oregon, and they said, 'I know how you can get rid of your invasive snails.' I said, 'No thanks. Figure out how to restore our natives, and then I'll talk to you about getting rid of our invasives.' I referred him to the South Florida Water Management District, because they've had trouble with the invasive snails in their STAs."

They sure have, I thought, remembering Eric Crawford's Snail-pocalypse.

Beck said that the reason that cattail always came back was the Army Corps of Engineers' policy of keeping the water deep in Lake Okeechobee to help with flood control. The deep water facilitates cattail taking over, which spurs the aquatic plant managers to hit it with herbicide and fire, rinse and repeat. One human impact has a way of cascading into more human impacts. Beck hands were tied. The snail kite coordinator for the FWC can only voice concerns and make recommendations to outside agencies, like the Corps, and then try to work with whatever happens.

In the inverted world that the snail kites now inhabit, two wrongs—deep water and killing everything—did seem to make a right, for snail kites at least.

Beck was optimistic that this habitat management approach could help grow the snail kite population. He said that snail kite numbers had been plateauing at about three thousand since 2017. Their population may not be able to grow further without gaining more suitable habitat. Beck said habitat managers have been trying to identify cattail monocultures in former snail kite areas. If they can kill the cattail and expand the amount of habitat available to kites, Beck said, perhaps the snail kite population could start to increase again.

To hear Beck tell it, the invasive plant managers seemed like allies rather than enemies, "us" rather than "them." It helped that there are now chemicals that can more narrowly target a particular plant without harming desired plants, and that dissipate quickly after the treatment.

I mentioned the periphyton wonderland that Paul Gray had shown me out at Indian Prairie. Beck said the marsh there was so open now because of a torpedo-grass treatment a few years ago. Score another one for invasive plant managers, I thought.

Beck cranked the airboat up again, and we flew over the marsh, through some small cattail stands, and around a large clump of willow. He slowed and cut the motor again in a pond ringed by willows.

"Just real quick," Beck said. He pointed to a tuft of willow. "Do you see it?" he asked. "A nest, right through there."

I grabbed my binoculars and scanned the willow branches. Finally, I discerned a collection of gray sticks.

My first snail kite nest!

Beck hopped up the steps behind our seats, which had been welded onto the cage that encloses the airboat's big fan. I loved that this airboat had been outfitted with a little platform specifically to observe snail kites. From behind his binoculars, Beck confirmed that there was a young kite in the nest. He came down and invited me to go up.

I could hear the chittering irritation call of the adult female kite. I didn't want to disturb her, but I was dying to see the young kite. On the top step, I gazed through my binoculars and spotted a fuzzy gray shape in the nest. A baby snail kite. It was exhilarating to see a little nestling with my own eyes, even if I could only make out a gray smudge amid the sticks. I wanted to linger, but I knew the female kite was eager for us to leave.

As we motored off, I tried to sear that fuzzy little snail kite into my memory. We stopped again a little later and talked surrounded by big yellow blooms waving in the air on two-foot-tall stalks, above perfectly round lily pads the size of dinner plates. The American lotus is in the Everglades, too, but not in the shallow marshes I spend time in. It was scenery befitting Alice in Wonderland. It was so hot that I could not tell if I was sweating or melting, but I wanted time to slow down. I wanted to stay in this beautiful marsh forever.

I was not expecting any bad news. But if I had to get bad news, it was as good a place as any.

At first, I didn't understand the significance. Beck simply said, "We're going to get a new lake regulation schedule, in the next six months or so."

I vaguely knew that the lake's regulation schedule is what determines how deep the lake can get, and for how long.

Beck explained, "The new schedule is going to hold the lake higher on average."

I was curious as to why the lake needed to be deeper, but I saved that question away for later.

He went on, "Anything above this level really starts to stress out the vegetation, particularly submerged aquatic vegetation. The water is already to the levee, so any water that's added to the lake beyond this point doesn't expand the marsh any. It just stacks water on top."

I thought about that. As you start to fill a bathtub, once the water reaches to all the edges, the water just gets deeper. I could see how it could become too deep to support submerged aquatic vegetation, since sunlight may not reach it in deeper water. Also, deeper water means bigger wave action, which can uproot vegetation.

"What about all of the wildlife that needs a shallow marsh?" I asked.

It seemed like Beck was used to the wildlife taking a backseat to human agendas, especially coming from outside agencies, and he took it with a mixture of dismay and acceptance. He said, "The regulation schedule weighed a lot of other interests other than the ecology of the lake. You know, water supply, flood control, navigation, and getting water south to the Everglades was part of that." He paused, as if choosing his words. But he just added, "This is the plan that they've settled on."

I said, bluntly, "I mean, it sounds like the worst plan for the marsh."

Beck said, "Yeah, but if you looked at all the alternatives. . . . They have these diagrams of like, okay, if you improve things for the ecology of the lake, all these other things go down. So, this plan is better for a lot of these other parts of the system."

Beck went on, "The lake's not going be able to have those dry times that really rejuvenate a lot of the vegetative communities and allow fire on the marsh, and those kinds of things."

I looked around us, trying to imagine deeper water. I realized that I hadn't seen a single wading bird since we arrived in the marsh. Wading birds don't like to get their bellies wet. They can't wade in three feet of water, much less five. I asked, "Would it be another two or three feet above this water level?"

Beck said, "At times, yes."

That was alarming. I couldn't help but think about Dr. Wiley Kitchens's words of warning about tweaking the system and getting a system-wide response that no one expects. The bigger the tweak, the greater the uncertainty. The cattail-moonscape cycle had become a predictable feature of the lake; would it work the same way with another two to three feet of water? Would the kites still benefit?

"Alright," Beck said, "I guess we'll turn the 'air conditioning' back on. That's the nice thing about an airboat."

Indeed, the wind in my face did feel great as we flew through the marsh toward the canal.

Tyler Beck exuded a determined forward momentum with his hand on the rudder and his boot on the accelerator. His eyes were inscrutable behind the blue shine of his sunglasses. I was grateful for his ability to strive for achievable wins and to stay engaged in a system that was far from an environmentalist's dream. Beck seemed to follow his own version of Jean Takekawa's mantra "There's always a solution." Working with what you see as the "least bad" is one way to persist in the face of what seems like overwhelming odds.

I now had a vivid picture of the snail kites profiting wildly from the cattail moonscape of today's Moonshine Bay. Watching so many of them flying around foraging was a thrill. The fact that snail kite advocate Dr. Paul Gray considered it a travesty told me that I was missing something important. I knew that as the Everglades science coordinator for Audubon Florida, Paul Gray has battled government agencies on behalf of the snail kites in Lake Okeechobee for many years. I resolved to follow up with him after I got home.

Weeks later, Beck emailed me to say that the fuzzy kite he had shown me in the willow nest had fledged. That year, the marshes of Lake Okeechobee fledged 122 snail kites, twice as many snail kites as the next highest site.

# 20 ARE SNAIL KITES TRASH BIRDS NOW?

Dr. Paul Gray of Audubon Florida shook his head ruefully when I told him of my adventures out at Moonshine Bay. He said, "The snail kite's a trash bird now."

That was unexpected. "Trash bird" usually means a species too common to excite birders, like pigeons. I knew Gray was as devoted as ever to snail kites, so I asked him to explain.

He painted a picture for me of three key moments in snail kite history in Moonshine Bay, illustrating three distinct versions of Lake Okeechobee's western marsh. Let's call them Moonshine Bay #1, #2, and #3. Through his vivid portrait of these three versions, I came to understand a key aspect of snail kite conservation that had eluded me until then. From this perspective, his "trash bird" comment and his anguish made sense.

Moonshine Bay #1 is the 1970s: This is the gold standard for Gray. He sent me a photo taken in the same area that is now subject to the cattail-moonscape cycle. In the photo I see shallow marsh to the horizon. There's sparse spikerush and abundant periphyton, much like today's Indian Prairie. In another photo, I could see the white blooms of fragrant water lilies spread across blue water with little clumps of saw grass. It looked like Grassy Waters Preserve, where I had recently seen so many clutches of native apple snail eggs from my kayak.

The abundance of native snails in Moonshine Bay made it such a crucial site for snail kite nesting that it was designated as a large part of the snail kites' Critical Habitat in 1977 (see figure 5.2).

What was good for native snails and snail kites was good for wading birds, migratory ducks, frogs, insects, and other animals. The bass fishing was legendary. It was a patchwork of shallow-water marsh habitats, floating leaf, gator flag, and native grass communities. It was a lot like the historical Everglades. Biodiversity was high.

The reason for this success was the water level schedule. During this time, Lake Okeechobee's water surface fluctuated from fifteen to twelve feet elevation above sea level, a measurement that describes how high the water surface sits, not how deep it is. This fifteen- to-twelve-foot fluctuation translated to just the right amount of water up on the marsh. At fifteen feet of elevation in the wet season, shallow water covered the marsh all the way to the Herbert Hoover Dike, with depths ranging from a few inches at the higher spots to several feet in the lower areas. This allowed for a wide range of vegetation types to thrive, including submerged vegetation. When it dropped to an elevation of twelve feet above sea level by the end of the dry season, the marsh became dry enough to regenerate. From the 1950s well into the 1970s, the lake levels oscillated within this optimal range, and the marsh and its wildlife thrived.

Then the policy changed. Starting in 1978 the lake schedule allowed the marshes to be above fifteen feet more than half of the time, and to regularly exceed sixteen feet. Over time, extreme water levels took their toll. Since the whole marsh is flooded when elevation of the water surface reaches fifteen feet, any additional water just stacks up on that. It was often too deep for wading birds. The marsh almost never dried, so seeds couldn't germinate; after a few years the submerged vegetation died. Deepwater plants died off and formed a ridge of dead plant material where the open water started.

In Lake Okeechobee's marshes, deeper water means more nutrient enrichment. This is because the higher the water levels, the more nutrient-enriched water from the deeper part of the lake washes up onto the marsh. The extra nutrients turbocharge cattail growth. This led to the disappearance of apple snails, wading birds, mammals, fish, and frogs. Moonshine Bay produced almost no snail kite fledglings after 1996. By the early 2000s, a cattail monoculture had closed over the marsh.

So that's Moonshine Bay version #2: miles and miles of dense cattail thanks to too-deep, nutrient-enriched water. No snail kites and extremely low biodiversity.

Gray said, "I was the first to tell the South Florida Water Management District Governing Board in 1999 or 2000 that kite nesting was virtually gone from the lake. They were meeting in Okeechobee, and it was a big surprise to them—no one had told them. That also was when the submerged aquatic vegetation had long ago disappeared, the bass fishery was in the tank, and everyone was mad."

But things didn't improve. The snail kite population declined in the early 2000s.

The reason for the deep water is simple: water supply. More storage capacity in the lake gives the South Florida Water Management District (SFWMD, or "the District") more flexibility in meeting the needs of agriculture and residential areas.

Of course, the Endangered Species Act (ESA) exists to stop government actions that could cause species to go extinct. It struck me as an obvious violation of the ESA to allow water levels that would prevent snail kites from using one of their most important nesting spots, in their officially designated Critical Habitat, especially at a time when their population was in a downward spiral. Apparently, the National Wildlife Federation (NWF) and the Florida Wildlife Foundation (FWF) thought so, too. In 2005 they sued the Army Corps of Engineers for jeopardizing the snail kites on Lake Okeechobee. They contended that maintaining elevated lake levels had "harmed the snail kite population by impairing their patterns of breeding, feeding and sheltering, in violation of the ESA." They sought to prevent the Corps from further damaging snail kite habitat with elevated lake levels.

Unfortunately, the lawsuit went nowhere. The last record I could find was a transfer of the case from Washington, DC, to a Florida court. I could not find anyone who knew how the twenty-year-old case was resolved, even at the NWF and the FWF. It seems to have simply disappeared. It shocked me that the snail kites could lose their nesting stronghold in Moonshine Bay with no consequence. Clearly, the ESA and the USFWS were nowhere near as powerful as I had imagined.

If Moonshine Bay #2 is the too-deep cautionary tale, one might hope that at least it would give the marsh some protection against droughts. Unfortunately, even when lake levels were allowed to be high, drought hit hard. Not only was there less rainfall hitting the lake, but it also received less inflow from rain-deprived areas upstream. And the District made it worse by increasing outflow. In the drought of 2001, when the water level dropped to the point where water could no longer drain out via gravity, the District did something new: They installed pumps to force more of the remaining water out of the lake. Knowing that they could install pumps, they could allow water levels to drop quickly. They installed pumps again in 2007, with similar results.

For the snail kites, water levels dropped too fast for them to even initiate nests in both 2001 and 2007. Snail kites had to disperse to whatever emergency wetlands they could find. It pained Gray to think of snail kites starving to death as the marsh dried down earlier than necessary. He said that people would ask him, "Where did the kites go?" And his answer was, "I don't know where they *went*, but they *ended up* in heaven." These two droughts combined dropped the snail kite population by 75 percent in less than a decade.

So, Moonshine Bay #3 is the bone-dry, drought-crisis version of #2. You could drive a field vehicle across the marsh. Wetland plants that normally grow submerged below the surface of the water were wilting in the sun. No snail kites.

Gray has lived through all of this history. When he first started working on Lake Okeechobee in the 1980s, it was a healthy, biodiverse, shallow marsh and a snail kite nesting supersite (version #1). He watched as year by year, too-deep water allowed a cattail monoculture to annihilate the marsh, making it unusable for snail kites or much of anything else (version #2). That was a huge loss for snail kites and countless other species, and Gray's protests made little difference. And, as if to add insult to injury, when the lake got too dry, the government sucked the last drops of water out from the marsh (version #3).

When the kites were going down with the ship, Gray tried in vain to get people to listen. It's even harder to sound the alarm now that kites are nesting there again, thanks to the cattail-moonscape cycle (which could be called version #4). Gray sighed. He said, "I am happy kites do well, but hardly anything else does. The snail kites live off this ecosystem that has been overrun by a cattail monoculture, followed by dead areas from spraying and burning, that stimulates an exotic snail, that the kites can thrive off. The ecosystem is trashed, and wetland scientists really don't want Lake O's marsh to be this way. These conditions eliminate too many plants and animals."

Although Gray often finds the invasive plant managers at the Florida Fish and Wildlife Conservation Commission (FWC) exasperating, when pressed, he doesn't really fault them for intervening in the cattail monocultures. He said, "I watch plant managers do what they can, and some good comes from it, but they can't fix the fundamental problem of gross mismanagement of the lake and its marshes, and the catastrophic loss of biodiversity that goes with it. That's what I am upset about."

And it drives him crazy when they tout it as a success. During one of our interviews, he told me that earlier that week, he had been to an FWC presentation boasting about the "success story" of Moonshine Bay, focusing only on snail kites. Gray said, "Based only on kites, it is a success. Kites are getting along great in the midst of what I see as an ecological disaster, and their success can lull people into thinking things are okay."

It did seem fishy to single out a species that depends on an invasive species, while most of the other wildlife that should be there is still gone. It didn't sit well with me that when an endangered species like the snail kite makes a big comeback because of an invasive species, it counts as a win for the US Fish and Wildlife Service. But when the snail kite has needed the Service to wield the ESA on their behalf, it rarely seems to happen.

Gray circled back to his "trash bird" comment. "If you go to the garbage dump, you will see thousands of birds: gulls, crows, eagles, and some ibis. Is a dump good habitat? That is more dramatic than marsh changes, but raises similar questions." To him, the state of Moonshine Bay is catastrophic. He said, "The lake has this big-time problem, and no one seems to care much." Gray wants to see it restored to the healthy shallow marsh it was back in the 1970s, Moonshine Bay #1. He wants to see abundant snail kites along with the native snails, fish, frogs, and wading birds that thrived there when the lake levels were more moderate.

It struck me that Paul Gray and Tyler Beck were both adaptable in their own way, even though they came to opposite conclusions about whether snail kites in Moonshine Bay counted as a success. I asked Gray if their different perspectives had to do with their distinct roles: The snail kite coordinator was squeezed between agencies, while someone from a nonprofit could push from outside.

Gray said, "I prioritize ecosystem integrity—to the extent we can get it—and agency managers have much narrower goals. If snail kites can live in Florida because we have an exotic apple snail, many will be satisfied. I won't be satisfied until they can live off the native snail in 'functioning' Florida ecosystems, and we are not there right now."

And now, he told me, the new lake schedule will keep it even deeper. The Herbert Hoover Dike, which contains the lake, underwent major repairs, allowing it to safely hold more water. He said bluntly, "The new schedule is going to keep the lake harmfully deep on a regular basis. It's likely going to be a disaster. The lake's already a wreck from the last schedule. They're like, 'Yeah, now we're going to make it even deeper.'"

This new schedule would be uncharted territory. I asked if it could produce unexpected results, throwing off the benefits that the snail kites have been getting from the cattail-moonscape cycle. Gray just waved a hand despairingly. "With snail kite numbers up, the Endangered Species Act isn't very useful as a cudgel," he said. "Their numbers are good. There's no leverage."

Gray said the real solution is to have more water storage and more water cleaning capacity (i.e., more Stormwater Treatment Areas, or STAs). If there were more of that upstream, the lake might not need to get so deep in the first place. And if there were more of that to the south, there would be more clean water available to the Everglades. That would be good for the Everglades and Florida Bay. And it would spare the estuaries on the east and west coasts from harmful discharges from Lake Okeechobee when its water levels need to drop. Gray was pleased that there are plans in the works for a massive new "EAA Reservoir" and an adjacent STA, in the Everglades Agricultural Area (EAA). He was quick to say that this will not be enough; the lake will still be kept too deep, and there will still be harmful discharges to the east and west. But it's a step in the right direction.

I also spoke with Jaclyn Lopez, the former Florida director for the Center for Biological Diversity. She mentioned the elephant in the room: The EAA demands to be protected from flooding, yet it also demands abundant water supply in the dry season, because that is the growing season. Lopez said, "If we wanted to talk about solutions, we have 540,000 acres in the Everglades Agricultural Area that are being artificially kept dry through canals and structures that are part of the historic Everglades 'River of Grass.'" She added, "If we want to point fingers and blame and get quick fixes, that's where you gotta go. That's that. All the rest of the ecosystems have nothing else to give. There's no more blood to give. The stone is dry."

Of course, the EAA is here to stay, ensuring uphill battles over water. Paul Gray seemed to have a lot of fight left in him. As much as he loves the snail kites, he would not let them be used as a distraction from the urgent need to protect Moonshine Bay as a vital habitat for a range of interconnected species.

I thought about the fuzzy little snail kite I had seen nesting in the moonscape of cattail stubs. It struck me that if snail kites at Moonshine Bay had become trash birds, it was in a cool, punk-rock way: flouting expectations, breaking rules, making something life-affirming with the broken pieces of their once glorious home.

Year after year, snail kites have somehow managed to take our environmental messes and make them into beautiful places full of hope and new life. And what a miracle that they do. It wasn't a promise for the future. And it wasn't absolution for our sins. But it was a glimmer of hope, that if we could start to restore Moonshine Bay and the Everglades, the snail kites may still be around to see it.

# 21 TWO HUNDRED SNAIL KITES

One mystery had been eating at me, and Moonshine Bay brought it to a head: Where was the Endangered Species Act (ESA)? Why did snail kites keep falling through its cracks?

It finally clicked after I dug into a controversy at Lake Okeechobee during the drought of 2011. It was a familiar showdown: human agendas vs. active snail kite nests. In February 2011, water levels were good, and many snail kites initiated nests. But the water started to recede too quickly as it flowed into canals to irrigate the Everglades Agricultural Area (EAA). On behalf of Audubon Florida, Dr. Paul Gray urged the South Florida Water Management District (SFWMD, or "the District") to implement greater water restrictions so that the nests could fledge their young. He got nowhere. In April, his pleas became more urgent but were still ignored.

In May, there were thirty-five active snail kite nests, and the water level continued to fall. The District planned to install pumps to extract more water out of the lake. The executive director of Audubon Florida, Eric Draper, wrote a letter to the District stating that pumping water out of the lake threatened active snail kite nests. It made no difference. On May 31, 2011, the District activated pumps and ran them for more than three weeks. More than half of the nests failed that year, partly due to the drastic water level declines in May, according to the UF monitoring report.

I simply could not comprehend how water could be removed from under the active nests of a critically endangered species, in its Critical Habitat, while its population had been reduced by 75 percent in the last decade. But that's what happened, with no intervention and no repercussions.

Finally, I found someone who knew exactly how something like this could happen. Steve Schubert, a wildlife biologist in the Vero Beach office of the US Fish and Wildlife Service (USFWS, or "the Service") retired in 2020 after twenty years. He became involved with snail kites in 2006 and wrote the 2018 biological opinion on snail kites at Lake Okeechobee.

First, Schubert answered my question: Why didn't USFWS step in? The prescribed role of the USFWS is to respond to, not initiate, interagency communication around harm to endangered species. Schubert explained that the ESA is implemented via a framework of interagency protocols. When the actions of an agency (e.g., Army Corps or the District), might harm a species listed under the ESA, that agency is supposed to officially consult with the Service. The Service does an evaluation, and if they determine that harm is likely, they respond with a formal statement called a "biological opinion" (BO).

In the biological opinion, the Service defines the type of harm (e.g., habitat loss or nest failure) and quantifies how much is likely to occur. The harm is collectively known as the "take." Hunting and fishing regulations had long used "take" to describe the legal capture or harvest of wildlife, and this was an expansion of the usage to include other kinds of harm. In addition to evaluating the type and amount of "take," the BO includes an assessment of jeopardy, as in, whether a proposed action would place the continued existence of a species in jeopardy. It is defined as a substantial increase in the risk of extinction, where the species' chance of "survival and recovery" is severely impaired. The ESA forbids government agencies from crossing that line. But that line can be very difficult to calculate.

Schubert said, "If an agency decides not to consult with the Service (i.e., not receive a biological opinion), the only thing the Service can do is communicate their concerns to the agency. The Endangered Species Act puts the responsibility on the agency (in this case, SFWMD) to do the right thing for the species. They have the freedom to choose their action, at their peril." The peril is that they could be sued under the ESA by a third party (often nongovernmental organizations). Even then, real consequences don't kick in until it goes to court and a judge makes a ruling.

I had thought of the Service as the bodyguards for endangered species. Talking to Schubert, I realized their authority is narrowly constrained, and much of what they do amounts to procedural compliance rather than substantive protection.

The failed snail kite nests in the drought of 2011 begged the question, Does the Service have *any* authority to constrain the actions of the District, and if so, when does that kick in?

By way of answer, Schubert shared with me a video of a SFWMD governing board meeting recorded in May 2017. On the surface, it looked like any government meeting, but when I realized what was being said, I sat up. Even though this meeting took place a few years after the controversial pump installations in 2011, the issue of snail kites on Lake Okeechobee was still hotly debated. In the video, a man wearing a pale-gray suit and a purple tie came to the podium and leaned on it with both hands. It was Peter Antonacci, the executive director of the District. He accused the Service of taking water supply decisions out of their hands. He concluded, "No fair-minded person can read the biological opinion any way other than the Fish and Wildlife Service is attempting to assert control over water supply during dry periods."

The person on the hotseat was Larry Williams, the head of the Service's Vero Beach field office, Schubert's boss. Wearing a dark-gray suit with a red-striped tie, he walked to the podium. He stood facing the nine members of the District's governing board up on a stage and tried to appease them as best he could.

One board member leaned into his microphone and asked Williams point-blank, "Are there *any* circumstances under which the Fish and Wildlife Service will act to *intervene* on the state's ability to allocate water?"

Williams reassured them that the take limit would not be a problem. If they needed a higher limit for take, he would provide it. He went further, adding, "The only situation where we wouldn't be able to provide additional take would be if the species was in jeopardy, which would mean there would be very, very few of them." He went on, "Right now we have about two thousand kites. And the 'jeopardy' level for kites would be something like two hundred."

I played that part of the video again, in disbelief. Two hundred snail kites?

The board member sternly reflected Williams's words back to him: "What I think I heard you say was, 'No, the Service will not *ever* act to intervene on the state's ability to allocate water. If we get close to take, the Service will add additional take. And the only time we wouldn't issue that is if we're at jeopardy, and right now we're at two thousand, and jeopardy would be under two hundred.' Do I have that right?"

Williams did not disagree. But somehow they still wanted the Service to cede more of its authority. The board member finally concluded, "You're going to be holding a single species in a very defined geographic area *against us.* It seems like all of the cards are stacked against us. I do think it's an infringement."

After more back-and-forth, Mr. Williams said, "There seems to be some discussion about 'constraints' that are established by the biological opinion. I would love to hear what those constraints are."

I would also love to hear what those constraints are. Because I wanted there to be constraints, and I wasn't hearing any. The Service was drawing a line in the sand at two hundred snail kites. As long as the kite population was above that, they would offer no resistance.

Schubert remembers being appalled to hear his boss cave like this. He said, "When Larry Williams said, 'less than two hundred snail kites,' did he do an assessment of the population? To me, it looks like he just made it up on the spot. Just because a person is in charge of an office or agency, that doesn't mean they understand the science of the decisions they are making. Or, worse yet, they do understand the science but still make a decision that impacts a species due to expediency or political pressure."

As frustrated as Schubert was, he understood the pressure that Williams was under. He said, "The Service managers in Atlanta wanted the Vero staff to iron out any controversies or problems between the Service and the Applicant (Corps or SFWMD)." Schubert said his supervisor told him, "Atlanta won't approve jeopardy opinions." It was an era of "go along to get along." Mr. Williams's bosses would not have his back if he tried to stand up to the District.

Schubert suspected that the higher-ups at the Service were themselves under pressure from Washington, DC. He said, "At that time, there was a lot of talk about Republicans wanting to water down the ESA. Their calling card was, 'No species has ever been recovered by the ESA.' Which, of course, is incorrect, but it's great political theater to congressmen and the public who are ill-informed." I had not considered the ESA itself could be at risk.

I was starting to realize that the teeth of the ESA came more from lawsuits against the Service or other agencies. But lawsuits are expensive and an uphill battle. In 2005, there were two failed lawsuits involving the snail kites: In addition to the National Wildlife Federation (NWF) suing over Lake Okeechobee, the Miccosukee Tribe sued over Water Conservation Area (WCA) 3A.

The "two hundred snail kite" meeting took place six years after the pumps were installed in 2011. Going into the 2011 nesting season there were 826 snail kites—a number snail kite advocates considered perilously low. I finally had an explanation for the lack of protection for their thirty-six nests. The number two hundred may not have been floating around yet, but it seems that regulators sensed that there was still plenty of "runway" in terms of time and kite population size, before intervention would be worth the political fallout. Given that political calculus, the outcome seemed so inevitable that I was impressed that Dr. Paul Gray of Audubon Florida had even tried to fight it. The same calculus seemed like it would play out in future droughts, too.

Still, one aspect continued to perplex me. I could understand how water allocations could tilt more in favor of agriculture than wildlife. But this seemed all-or-nothing. Agriculture got all it wanted, and I could see no sign that wildlife was considered at all. I pressed Gray, "Doesn't agriculture have to be rationed during a drought year?"

Gray explained that agriculture does get rationed, but the math is funny. The District guarantees water supply to an area which includes the industrial-scale farms like those of Florida Crystals and US Sugar (aka "Big Sugar") in the Everglades Agricultural Area (EAA). And that guarantee is in place regardless of how low the lake level drops. They get a *bigger* allocation on drought years, so much bigger that when the 40 percent ration is applied, they still end up getting *more* water in drought years than regular years. It was like a CEO taking a comparatively small pay cut right after giving themselves a large raise, so they still end up ahead.

Gray added, "While it makes some sense to give a plant that didn't get rain a little more water, the agencies also have a responsibility to protect wetland ecosystems from excessive drought stress. Giving away more water makes damage from the drought even worse."

What really gets to Gray is that it's not even clear that all of the extra water is necessary to protect agriculture. Gray found that sugarcane crop yields showed no impact from the droughts, and that they didn't even use all the water that was made available to them. He laid this out to the District, urging them to do their own assessment of water allocations, droughts, and crop yields, to avoid sacrificing wildlife to supply water that was not needed.

He got no response. I wondered how Gray kept getting up to fight another day.

The truth is, the District is hemmed in as well. Big Sugar is extremely generous with political donations. Stakeholders with more money and power can always bully agencies into giving them more than their fair share, at all levels of government. PACs tied to Florida Crystals reportedly gave $1 million to the MAGA campaign in 2024.

From the outside, it's easy to think I'd have made different choices if I were in the Service or the District. But time, resources, and political capital are limited, so picking battles is important. Steve Schubert drew a line between ground-level staff and higher-ups like Antonacci. He said, "Other agency personnel at the technical level were usually intelligent, honest, and easy to get along with. It was just that their managers sometimes had different ideas. And I believe there was fear in the SFWMD that any employee who did not tow the party line would get fired. So that had to be hard for some staff there."

Paul Gray echoed this. He said that on multiple occasions, with multiple agencies, he has argued with an agency person in a meeting, after which the person confided, "Actually I agree with you, but I have to follow the agency position." Once, he was talking with someone about an issue, and they said, "Look, I lost that argument in the executive director's office. So don't blame me." Audubon pushback helps keep ecology on the table—but only to a point. Special interests can indirectly force the hand of agencies, and it's the agency staff that take the heat. Sometimes the public pressure that Audubon applies in the media can give agencies political cover to make concessions. Lawsuits can be effective, too, when they are successful.

Many consider the ESA to be our nation's most powerful environmental law. I realized that I had to accept the extent to which even the ESA can be blunted by money, power, and bureaucracy. The snail kites deserve more protection than they get from government agencies, but it is not the fault of the ESA itself. Schubert told me that full implementation of the ESA would require more funding, enough skilled, well-trained, and conscientious staff, and supportive management in Florida and the regional office in Atlanta.

I now understood better how the ESA is implemented for endangered species generally, and the snail kite in particular. Snail kites have paid a steep price for our choices, but the species has so far managed to survive in spite of us. They've been bold and adaptive, responding quickly to changing conditions and seizing unexpected opportunities. As a result, their population has more than quadrupled since hitting a low of just seven hundred birds in 2007. Ironically, becoming what Paul Gray calls a "trash bird" has helped them more than their endangered status ever did.

# MIRACLE ROOST, REVISITED

# 22

One morning, as I was digging into the scientific literature on snail kites, I was taken aback by an unexpected reference to the 372-kite roost Jean Takekawa had seen in 1985 at the landfill site. I sat up in my chair and started skimming frantically and looking elsewhere for confirmation. Was it possible?

It was! Takekawa's miracle roost was still there! I could hardly believe it. I immediately knew I needed to see it for myself. I suspected that, like Grassy Waters Preserve, it would provide a piece of the puzzle I didn't even know I needed.

On a hot July evening, I found myself up on a metal tower with David Broten, the environmental programs manager for the Solid Waste Authority (SWA) of Palm Beach County. The property is east of Lake Okeechobee, and adjacent to the southeast boundary of Grassy Waters (see figure 1.2 and figure 7.2). Broten was tall with close-cropped salt-and-pepper hair and a pale-blue polo shirt. Mandy Krupa was up there with us as well; she had blue-rimmed glasses and a blonde braid. Next to me was my former student Zoe Sabadish, a recent graduate. She had brown curly hair and carried a camera rather than binoculars.

When we got our first look from the top of the tower, Zoe and I gasped. Zoe said, "It's like stepping from one world to another." The forested islands below hosted a cacophony of baby birds. Jean Takekawa had mentioned that the roost was amid a wading bird colony; I had not imagined how busy and overwhelming it would be to behold. There must have been a thousand birds, or more. A glossy ibis nest right next to a wood stork nest, and then a roseate spoonbill one, and it just kept going: white ibises, cattle egrets, anhingas, little blue herons, tricolored herons, great egrets, snowy egrets, black-crowned night herons, green herons, great blue herons, and more.

Three young roseate spoonbills, pale pink and adult-sized, were crowding around their parent at the end of a branch. The bright-pink adult inserted its bill into a baby's open bill while the siblings clamored for their turn. Everywhere I looked, pairs or trios of goofy young birds were either flapping around as they received their last feeding before bedtime or stoically waiting for it.

Three wood storks, as large as adults but with light fuzz on their heads, stood on their nest, their bills tucked into their shoulders. The middle one was facing the other way.

Krupa laughed, saying, "If the middle stork turns around, it will knock the other ones off the nest!"

Zoe said, "I've never seen birds like this. It's like a city. All walks of life coming to a place of rest." The website for this part of the SWA property reads, "In 1985, this area was permanently designated as a conservation area, as it's a vital habitat for the endangered snail kite." That was Jean Takekawa's legacy, I thought. I wondered how many snail kites had survived drought there, and how many tens of thousands of wading birds had been born there in the intervening forty years.

"It makes me think of all the magic places that are probably lying beside the highly developed ones we so often interact with," Zoe said.

Near us was a 150-foot-high grassy hill with a broad, flat top, which in flat-as-a-pancake South Florida can only be a landfill. A loud industrial hum from the adjacent waste-to-energy plant competed with the squawking birds.

The county had needed a permit to construct the landfill because the Clean Water Act of 1972 protected wetlands from being dredged or filled in. And it turns out, when a wetland hosts an endangered species, the Endangered Species Act (ESA) can come into play with these permits. Takekawa's successor at the US Fish and Wildlife Service ensured that the permit included two requirements: protecting the roost site and close monitoring the snail kites at the site for the first seven years.

The county hired a biologist named Dr. Darren Rumbold to head up kite monitoring at the roost. When I spoke to him by phone, Rumbold told me he had overlapped with Takekawa for a brief time at the Loxahatchee National Wildlife Refuge. Krupa was one of his kite-counters starting in the late 1980s. Rumbold installed four observation towers, so the kite counting team could observe the roost from every vantage point. People in each of the towers communicated by radio to ensure they did not double count.

Rumbold wrote a paper on the monitoring project, which is how I found out the roost had not been destroyed. Rumbold told me his most vivid memory was when a bald eagle flew in, and five thousand wading birds instantly took to the air. He was also amazed at the 212 snail kites roosting there during a drought in May 1989, which he had seen along with Krupa and another dedicated kite-counter named Mary Beth Morrison. Those 212 kites represented half of the estimated snail kite population the previous winter. Rumbold noticed that when the snail kites came in to roost, they were almost always coming from the direction of Grassy Waters. Same as in Jean Takekawa's day.

I asked Broten and Krupa, "Have you been seeing snail kites out here recently? I mean, it's not a drought year."

Broten chuckled. "Oh yes," he said. "There were sixteen last month. The monthly average is thirty snail kites." I had read that most snail kite roosts are only used for a few years before being abandoned. This one has now been in continuous use for several decades, even on wet years.

I asked Broten what snail kites looked like when they were roosting. I imagined them all arrayed like the wading birds below. Could he see the snail kites tuck their bill into their feathers to go to sleep?

With another good-natured chuckle, Broten said, "No, not at all. They fly into the trees and out of sight, until they fly out again in the morning. The roost count is all about spotting them as they fly in."

Looking out at the many forested islands, I asked, "Do you think the kites are roosting in the same exact spot as the 372-kite-roost, from 1985? Or is it even possible to tell?"

Broten gave a big smile. "Oh, yes. It's the same spot. There's no mistaking it. You'll see." Broten pointed out a prominent palm tree reaching above the other trees in the island directly across the pond from us. Between the palm fronds, I could see an ibis in a nest. It would be a crowded roost if they all landed in that one palm tree.

We had started two hours before sunset, well before the snail kites would start coming in for the night. We kept our eyes peeled for them.

A dozen or more white ibis appeared over the landfill and flew into the rookery. Three wood storks became visible above the landfill and flew our way. It almost seemed like they had been foraging there, and I asked Broten about this.

"Oh, definitely," Broten said. He explained that the part of the landfill we could see was no longer being used, but on the north side there was a "working face" the size of a basketball court. It gets covered each day with mulch, soil, and ash, but in the meantime, ibis forage amid the exposed food waste. Broten said that when the SWA shifted to waste-to-energy, they wondered if some of the birds would leave if there was less garbage landfilled. Even though the wood storks rarely foraged in the garbage, they monitored wood stork nesting and were relieved to see that it did not cause any problems. Wood stork numbers were similar before and after the change; if anything, more moved in.

"How strange," I said. "That exposed garbage could become a valuable asset to an endangered species."

"I know," he said with a big smile. "It's crazy."

Snail kites don't forage in the landfill. There was initially a concern that their prime foraging grounds in Grassy Waters could be affected by the trash incineration plant built on the SWA property. Emissions have the potential to contaminate nearby wetlands with various toxins, including mercury. Rumbold investigated this as part of his surveying years ago. To get snails for testing, he kayaked out in Grassy Waters in the middle of the night so he could catch them laying their eggs on plant stems. To his surprise, toxin levels were not worse than other places in South Florida. Although the required seven years of monitoring elapsed long ago, the SWA still closely monitors the snail kites.

I was amazed that in the time since Takekawa had witnessed the roost, an entire solid waste complex had been built around it without diminishing the wildlife resource one bit. In fact, given that we have made drought more severe, the kites need artificially deep roosts now more than ever. A gator slowly glided across the wide expanse of open water. Krupa said that she thought the gators were a big reason this was such a desirable spot for nesting and roosting. It's the same reason that snail kites want at least 1 meter of water under their nests: They want it to be deep enough to attract gators, whose presence deters predators like raccoons. Ten-foot-deep mining pits provided that. It reminded me of children's book illustrations of castles protected by moats with alligators.

And shell pits with gator-deep water weren't the only human impact that had inadvertently become a benefit.

"Is that all Brazilian pepper?" I asked.

Broten nodded and said, "Actually, the Fish and Wildlife Service forbade us to do anything to the Brazilian pepper or face a fine." He laughed and shrugged. "It's crazy, right? But it makes sense. We weren't allowed to modify the shell pits, either."

People imported so-called Florida holly in the twentieth century for its bright-red berries. Brazilian pepper went on to become one of Florida's most pervasive invasive plants, as it grows rapidly and is almost impossible to get rid of. I had lost many battles with Brazilian pepper in my own yard and on campus. I never imagined this invasive plant could become a conservation asset. But here we were. I couldn't help laughing with Broten.

Broten said, "There's our first kite!" I followed his pointing finger, and there it was. The white band on its tail was prominent against the darkening trees near our horizon. The kite was coming from our right, the direction of Grassy Waters. It flew behind the tall palm tree and disappeared. Over the next hour, another followed.

A sprinkle turned into a downpour. The four of us all laughed and huddled under the metal roof of the viewing tower, as rain sprayed in through the open sides. Cool water sprayed on my face and made my shirt stick to my arms. The young wood storks and other babies stood on their nests in the rain as stoic as ever.

The rain dissipated. A rainbow appeared just to the left of the cabbage palm. A large cloud to our right glowed creamy yellow with the setting sun. Another snail kite soared against a golden cloud high above us and then circled the roost before vanishing into the foliage.

Krupa noted that the kites may have been sheltering somewhere, and now maybe there would be some final kites coming in. The yellow cloud now glowed orange, the last hurrah of the sunset, as the roost began to fade into the dusk. We watched as another kite flew over the pond and into the roost.

And one more.

This seemed like the right time. I said, "Can I read you all something that Jean Takekawa wrote?"

Everyone turned my way, their eyes still on the sky. I read from an email Takekawa had written to me: "I really hope you get to go on a count, that would be kind of full circle for me! If you do, please thank the staff involved for their efforts to continue to provide that urban oasis, so important to have those in south Florida. It gives me more hope that it is still there."

Krupa broke into a big smile. "That gives me goose bumps!" she said, adding after a pause, "Tell her it was an honor to be part of this."

We all watched in silence as a final snail kite flew across the golden clouds into the roost.

It was a proud little moment for humanity. As a species, we have done a lot of damage to the environment. Before me was Exhibit A: busted-up bedrock, invasive plants, a loud industrial hum. How strange that this very wetland was also one of the few things I could point to that people had done to help the snail kites. The life force of this wetland throbbed with the little bird families. It still welcomed the kites in to roost. The snail kites were showing me that under certain conditions, industry and nature could coexist and even thrive together.

The snail kites' flight path that evening connected the two Floridas. By day, the snail kites foraged in a piece of the historical Everglades, Grassy Waters. And by night, they sought safety in this deep scar on the landscape. I recalled that this "novel ecosystem" was within the footprint of the original Loxahatchee Slough, an eastern arm of the original Everglades watershed (see figure 5.1). Takekawa had surmised that snail kites may have been coming to this area for thousands of years. They had found continuity in the disruption. I realized that the snail kite had become a living bridge between two Floridas: the one that existed for thousands of years, and the human-dominated one that they have adapted to use.

Thirty minutes after sunset, Broten closed his yellow field book with a satisfying crack and said, "That's a wrap!"

As our feet chimed on the metal steps, Krupa said, "It's a shame there weren't more kites for you two to see. There were more last month!"

"I'm thrilled to have seen kites fly in to roost here, and to be here with you," I replied. The roost was there for the kites when they needed it, and since the rains had returned in the last few days, most of the kites have moved back to other wetlands.

Zoe chimed in, "It makes me think of all the magic places that are probably lying beside the highly developed ones we so often interact with." When, on the drive home, I asked Zoe what she made of it all, she said, "I think that ecosystems change beyond the kites' control, and the birds change with them. That's what life does. It adapts. It becomes something new. Nothing stays the same."

It struck me that this kind of radical acceptance was exactly what snail kites needed from us right now, as we try to bridge the past and the future. Zoe Sabadish and Kait Kennedy are part of a new generation of conservationists. They are ready to meet the challenges of their time, just like Jean Takekawa did in hers.

# GIDDY UP ON THE SNAIL KITE RANCH

# 23

Sitting in the saddle with the leather reins in my right hand, I took in my horse. Abundant white hairs stood out against her rust-colored mane and coat. It turned out that her name, Roany, is a term to describe that coloration.

It was obvious that Roany had clocked me as an amateur right away. Every time my attention strayed to one of the numerous snail kites flying past us, she would stop and duck her head down to munch on the lush greenery at her feet. I would notice and get her going again. She would slosh through the water with a lazy grace. Then the next kite would fly past us, and we would repeat the cycle.

I was at Pearce Ranch, southwest of Lake Okeechobee (see figure 1.2 for location). The ditch along the farm road was dotted with little willow trees, and almost all of them bore the silhouette of a snail kite. They flew over the canal and over the marsh, staring down, searching for snails. The water was clear and about two feet deep, and the grass was spaced well enough for them to spot snails. There were certainly plenty of snails. Every willow's trunk was studded with clusters of pink eggs. Empty shells of gold and brown littered the edges of the canal and the little stumps of dry land across the pasture. Near some distant willows, two snail kites flew in tight circles around each other. I wondered if they were courting, especially since that very swath of larger willows was where they had nested the previous year.

I counted a dozen species of birds, including wood storks, roseate spoonbills, great blue herons, white ibises, glossy ibises, great egrets, cattle egrets, limpkins, and black-necked stilts. There were even green herons and anhingas perched on branches of willow over the deeper water. The air was aswirl with more dragonflies than I had ever seen. A bright-scarlet one hovered near me.

FIGURE 23.1. A horse named Roany at Pearce Ranch gives the author a ride across flooded pasture where cattle and snail kites happily coexist. Photo by Hilary Flower.

It was hard to imagine that just one month prior, the field had been parched, dusty, and dead. It had been the fifth-driest May on record.

I couldn't help but wonder, Could recovering an ecosystem be as simple as, "Add water"? It wasn't an ordinary ecosystem, that was for sure. But the wildlife was embracing it wholeheartedly. I thought about how this spot would have been Everglades sheet flow at one time. Now it was recovering many aspects of that ecosystem, without losing the ditches, fences, and cattle. As I tried to imagine the snail kites' future, I was starting to realize that at least some snail kites might ride off into the sunset on a ranch like this.

Every time my horse stopped, Matt Pearce got farther ahead of me. From his cowboy hat to his well-used leather boots with spurs, he looked like he was one with his horse. Pearce was whistling and cajoling a few dozen brown cattle around a fence, so they could graze another part of the impoundment.

Cattle ranching was central to Florida's early history as a state, and it's still going strong in the interior. Pearce told me that he was an eighth-generation Florida cattle rancher. He proudly reported that his son had just sold ninety calves that morning.

The cattle, Pearce had told me, were part of what made the ranch so good for snail kites. Without cattle, the greenery would grow in too thick and tall. The cattle also aerated the soil and brought out crustaceans and invertebrates for the wading birds to eat.

Pearce's friend Travis Thompson had told me over the phone that when snail kites arrived at the ranch the year before, they quickly adapted their foraging behavior. If a tractor went through the flooded pasture, snail kites would come down and take the snails stirred up in its path. Similarly, when the hunters put up duck blinds, snail kites would sit on the blinds or in nearby trees, ready to snatch snails disturbed by the hunters wading out to retrieve fallen ducks. Once again, the kites had found an angle no one anticipated.

I watched as one by one, the cattle rounded the bend and started to run on the other side of the fence. Water splashed up at their feet, making it look like they wore big silver crowns for shoes. The cattle, too, were adaptable in surprising ways, completely at home in this flooded landscape.

Pearce closed the gate and waited for Roany and me to catch up. We had to go through another ditch, and the horses were not thrilled to trudge belly-deep. I remarked that at least there weren't any gators in those ditches, given that it had so recently been a dry pasture. He shook his head gravely, saying that there were gators all along the ditches. He had lost more than one dog over the years; an impulsive splash into the water on a hot day, and that was that. Ouch. The wry adage was proving true: "To check if a waterbody in Florida has a gator in it, reach down and touch the water. If it's wet, there's a gator in there." Two dogs accompanied us that day to help herd the cattle. They stayed close to Pearce as he crossed the ditches.

The cattle that Pearce had just moved from one field to the other included many awkward, leggy calves. I asked, "I'm assuming the calves are too big to be taken by gators?"

"No, unfortunately," he replied. "We had to put a calf down just this week. Half of its leg had been bitten off."

Ouch again. Ranching in gator country was not for the faint of heart, I thought.

Pearce told me that no water drained off the land; it stayed contained. And the field naturally tilted very slightly to the west. He said that years ago previous owners had laser-leveled it to grow tomatoes and bell peppers. So, when summer rains turn it into a wetland, it produces very gentle sheet flow, a rare thing in modern Florida. There were several other snail kite–friendly impoundments within the ranch.

Rainwater stays on the land, because the US Department of Agriculture, Natural Resources Conservation Service (USDA, NRCS) bought the rights to use it as part of a twelve-thousand-acre Wetland Reserve Easement. The purpose of the easement is to store water and benefit wildlife.

Privately held land has played an ever-more-important role in conservation efforts in recent years, and the snail kites have been among the beneficiaries. Pearce explained that there are three layers of rights to the land that we were on. The base layer is the owner, the Zipperer family. They sold the *use* rights to the government for millions of dollars. Above that is the Pearce family, who lease grazing and hunting rights from the Zipperers. The Pearces use it for raising seven hundred head of cattle, a duck hunting outfitter service, and other activities like ecotours. He says that within this pecking order, raising cattle and his family's other activities are a privilege, not a right.

Pearce said that it's common for flooded pastures like this across the interior of the state to be visited by wildlife, including listed species like the snail kites. But often the instinct of a ranchers is to keep their mouth shut, being wary of outsiders who do not understand ranching, and who may come in and start dictating what can and can't happen. Pearce acknowledged that it would be easier to just take the boards out that hold the water back, drain the pasture, and say goodbye to the endangered species.

Pearce says that, while it's true that anything he wants to do has to be approved by the NRCS, it's a two-way-street. The NRCS has helped fund roller-chopping of invasive woody vegetation. Pearce said that when he first arrived five years prior, the Brazilian pepper and primrose willow were so tall and thick that you couldn't see across the field we were on. Now it's entirely open, and native plants are popping up in patches. The NRCS also helped him install the fence that he was herding the cattle around. Rotational grazing allows Pearce to prevent any one pasture from becoming overgrazed.

It's an amazing success story for government-private partnerships and endangered species. This was not lost on the NRCS; they made a documentary about their partnership with Pearce that focuses on the benefits to snail kites, and they show it around the country.

The previous August, when Pearce first noticed snail kites along the canal, he had pulled his truck over. He grabbed the handheld counter he uses to count cattle. He told me that the hair stood up on the back of his neck as he counted more than fifty snail kites. He called up the agencies, and the UF monitoring team started coming out to do nest surveys.

On his phone, he showed me a video he had taken: A woman is above her waist in a canal, raising a nesting pole up to a nest in willow. Pearce said that he had turned to his duck guide and asked, "If a gator attacks her, do we go in to save her or the birds?"

His guide responded, "The birds."

Pearce worked with the UF researchers to monitor the snail kite nests and band the young. He brought his daughter in for the banding sometimes. Holding the young kites got him choked up. The ranch ended up hosting the second-highest number of snail kite nests that year (2023), fledging more than sixty-six kites. Pearce said there were many more snail kite nests that they didn't count because they couldn't get to them for monitoring. The nest success rate was 68 percent, which is quite high for kites. And it had all happened in August, which is when I had seen a juvenile snail kite in a nest in Moonshine Bay with Tyler Beck of FWC.

Pearce was all in. He went to the annual UF Water Symposium and spoke on panels with scientists and agency folks every year. I've given talks there about the Everglades and climate change. Little did I know that this rancher was in another room charting a new vision for Florida.

I was glad that I had gotten to see snail kites in the Everglades before they left. Pursuing the snail kites over the last year had brought me into one unfamiliar landscape after another in parts of the state I'd never been to, bursting with wildlife. This was my first time out on a ranch in Florida, and it had certainly opened my eyes. On the gate leading into the ranch, I saw a sticker that read, "Cows keep Florida green." With few exceptions the alternative to ranches is not a return to the Everglades; it's either high-intensity farming or residential developments. Here was a working ranch that doubled as an impromptu wildlife sanctuary.

FIGURE 23.2. Two baby snail kites (and a third lower down on the left) in a willow nest at Pearce Ranch; this working cattle ranch contained the second-largest snail kite nesting spot in Florida in 2023. Photo by RC Gilliland/All Florida (allfla.org).

Clearly, being a wild animal in Florida had changed in ways that I had not fully registered when I was alternating between my urban life and the Everglades National Park. Most of the emergency wetlands that Jean Takekawa had discovered in the 1980s had been ditches, retention ponds, and farm canals. In the last two decades, snail kites had established nesting supersites in hidden wetlands in the human landscape, foraging for snails from the hydrilla covering the deep part of Lake Toho in the early 2000s, stormwater treatment areas, the cattail-moonscapes of Moonshine Bay, and, apparently, cattle ranches. Because of the scarcity of the native apple snail, I realized, novel ecosystems are almost all they have.

I thought about how the C-44 Stormwater Treatment Area (STA) had become a snail kite nesting supersite in 2021, fledging 72 snail kites. In both cases, it was because flooding a pasture produced a boom in apple snails.

This got me thinking . . . What if there were lots of ranches like this, where that switch could be flipped every year, drying down in between to prevent the aquatic predators from building up?

We dismounted from the horses, and Pearce loaded them back into the trailer with the dogs. As we sat together in his truck on our way back to the gate, I wondered how far we could push the "snail kite ranch" conservation strategy. Could it provide a snail oasis even in drought years?

I asked, "In a drought year, would there still be water in that canal? Would there be a way for you to keep this impoundment flooded?"

Pearce said that yes, there would indeed be water in the big canal between the county road and the eastern end of the property. With help from the NRCS, he could install a pump system that could keep the impoundment flooded even in droughts. One snail kite researcher in the 1980s, Dr. Paul Sykes Jr. had urged the creation and maintenance of a number of "selected 'islands' of [snail kite] habitat scattered within the historic range" for use on drought years, and they should be "as large as practical, and none less than 40 hectares." When I spoke with Sykes by phone, he considered that the STAs to fill that role now, but no cell is being optimized for snail kites. Critics of constructing a tailor-made snail kite habitat argue that it's impossible to predict the snail kites' movements on any given year, so it would be a fool's errand.

Even if it were successful, some would dismiss such a heavily engineered approach to wildlife management as little more than a petting zoo. With so much wildlife banded and tracked, we know where many individuals were born, who they're mating with, and even the length of their leg bones. If we create a series of wetlands engineered to optimize conditions for a single species, it can sound like cheating. Still, the fact is that natural ecosystems untouched by humans are long gone in Florida. We have entered a time of engineered ecosystems and accidental ones. Even Grassy Waters Preserve is subject to drinking water extraction, and the Everglades National Park is ultimately shaped by water control structures upstream, which determine how much water moves through it, where, and when.

Given that Pearce Ranch alone produced almost half of the fledges in a single year, wouldn't it be worth a try? I felt confident that no such project would be forthcoming, but I wished it were on the table.

Pearce stopped the truck to point to a nearby snail kite flying about twenty feet from the truck. "Look how bright their colors are," he remarked. I never would have thought to call the male snail kites colorful, but it was true: The bar across its tail flashed electric white in the sunlight, in dramatic contrast to its slate-blue body, the orange of its legs and the base of its bill, and the shocking red of its eye.

# THE SNAIL WHISPERER

# 24

Someone was waving at me. I strode over the dry grass and into the waters of the marsh. Dr. Phil Darby gazed at me over reading glasses as I climbed onto the airboat to shake hands. He looked equally ready to stand at a podium or jump into the marsh.

Darby and his field tech Adam Griffith had driven the eight hours down from Pensacola to Lake Kissimmee in part to monitor a population of native snails that he had found there the year before. Lake Kissimmee is in central Florida, about twenty miles south of Lake Toho, and sixty miles north of Lake Okeechobee (see figure 1.2).

Were the native snails staging a comeback? I had come to see it for myself.

And I had one other goal for the day. I hoped to get an answer to a central mystery: What had caused the native apple snails to become so scarce in the Everglades? And were they the key to the snail kites' future? I hesitated to lay at Darby's feet the entire mystery surrounding the native apple snail and the Everglades. But all my other experts had said, "Ask Dr. Phil Darby."

So that was what I was there to do.

I am not sure what surprised me more, that Darby could so easily answer my burning questions, or that he would cause me to see the native snail in an entirely new light.

Darby often brought up his mentor and hero in this work, Dr. Rob Bennetts. Bennetts, who died in 2022, had been at UF when he had first arrived in the early 1990s. Bennetts was the one who first introduced Darby to the Everglades, apple snails, and kites. It seemed to me that for Darby, Bennetts was always with him in spirit when he came out to the Everglades. Darby had shifted from kites to snails when he realized that prey were the biggest factor for snail kite survival.

Darby sat on a high seat with the controls, and Griffith and I shared a bench at his feet. We all donned headphones, he cranked up the motor, and we were off. Patches of lily pads became visible way ahead, whooshed toward us, and disappeared behind us. We passed a small hummock occupied by a common gallinule and her fuzzy gray babies. Countless patches of lily pads appeared and flew past.

We slowed as we approached a large cluster of bulrush between us and the open water. Darby cut the motor. Long green stems of bulrush came two to three feet out of the water, topped by rust-red tassels.

I told Darby that I had never seen bulrush so sparse.

Motioning with both hands, he said that he had seen snail kites come down and pluck snails out of the water right here. The stems were wide apart enough that the kites could fit their wings in there.

Darby reached down to grab a bulrush stem with a trail of a dozen native snail eggs, and snapped the stem below the eggs. He reached down for a large lily pad, carefully folded it around stem and eggs, then gently fit the little green package into an old shoe for further protection. He tucked the shoe in a safe place below the seats. He wanted to take the eggs back to his lab to hatch.

He told me how thrilled he had been the previous year to find "native snails all over the place" among the bulrush here. He said the vegetation was right, the substrate was right, and the native snails were abundant. I flashed to my recent thrill of seeing a ton of native apple snail clutches while kayaking in Grassy Waters Preserve. Another successful native apple snail habitat could hold big lessons for Everglades restoration, and this one had more potential for expansion.

He had reached out to FWC's snail kite coordinator, Tyler Beck, saying, "Y'all did something right on the other side of Ox Island!" Now he had come back to sample quantitatively. He was hoping that FWC might take management lessons from this successful spot.

Lake Kissimmee was one of the two lakes where Steve Beissinger had found nesting snail kites in the drought of 1982. Before the drainage and development of the twentieth century, Lake Kissimmee had been part of the continuous flow of water in the Everglades watershed. It's part of the snail kites' birthright.

In all my confusion and distress over the population crash of the native apple snails, I had never stopped to consider that their comeback may already be under way, until Darby had told me of his discovery. One of the questions I had stored up for Darby was if the native snails themselves might be endangered. It seemed like so little money was invested in studying them that they could be endangered without us knowing. Many of the snail kite experts I had spoken to considered this a real possibility.

Darby said, "Functionally, yes. Actually, no. There are probably hundreds of millions, if not a billion native snails in peninsular Florida. Functionally, though, yes, because their densities are so low. The animals that have relied on them as prey, going back to the evolution of the snail kite and limpkins and other things that eat snails, these predators are getting little benefit from the native snail now. Even the snail kite, the species that evolved to target the native snail, doesn't use many of them anymore to support reproduction, making kite babies."

I said, "So for snail kites, the native snails may as well be extinct. But in terms of their population size, native snails are not in danger of truly going extinct?"

Darby nodded. That was good news.

I launched my next volley: "What caused the native snails to decline? Did the invasive snail cause it?" It was a dynamic playing out around the world: When invasive species came in, native species started tanking. Plus, people often pointed out that the decline of the native snail coincided closely with the invasive snail becoming prominent. The invasive snail seemed like the obvious culprit, and we just needed to know how.

Darby ticked through all the possible ways an invasive snail could potentially harm the native snail. First, parasites: The invasive snail could have arrived in Florida carrying a parasite deadly to the native snails. He told me that he had investigated this but hasn't published his findings yet. He had teamed up with Dr. Tim Collins and Dr. Paul Sharp from Florida International University (FIU), and they did a full parasite inventory on more than a hundred native and invasive snails. The native snails did sometimes have lung flukes. These parasites are harmless to the snails but were known to have killed a couple of kites who had eaten infected snails.

"What about the invasive snails?" I asked. "Did they have parasites?" I was already flinching. If the native snails had parasites, surely the invasive ones were loaded.

Darby shook his head. He said, "We didn't find one single fricking parasite. I mean zero."

Of course, absence of evidence is not evidence of absence. But this study certainly threw cold water on the parasite hypothesis.

Darby climbed back up into his high seat and we slowly motored around the patch of bulrush looking for more white eggs.

A large bird flying above with a dark body and wings, and a bright-white head and tail, caught my eye. It soared in circles above us. I pointed it out. He cut the motor and said, "I can never get tired of seeing a bald eagle." Darby's field tech, Griffith, pointed out that a juvenile bald eagle was soaring behind the adult. It appeared the same as the adult except for its dark head and tail. After a while, they flew into a nest in a large cypress tree on a nearby patch of land. Darby said that was Ox Island. Lake Kissimmee is so large that the islands within it can look like the lake shore.

"What about competition?" I asked. "Did the invasive snails outcompete the native snails for habitat?" The two species were not often found in the same place. It made it seem like the invasive snail had pushed the native snails out.

"There's a lot of speculation about different ways that the non-native snail outcompetes the native," Darby replied. He pointed out that if the invasive snail were outcompeting the native snail for food supplies, you would expect to see that reflected in the habitat. He noted that periphyton was still plentiful in the Everglades.

Good point, I thought. So, food competition was out.

The snail eggs we'd recently seen reminded me of reports from Asia where a different apple snail species had preyed on the eggs of native snails of the area. I asked him if that could be happening here. Darby said there weren't any indications of snail-to-snail predation. "Plus," he said, "we do see coexisting populations of native and non-native snails. We've seen ten years of data where we go back to the same places in Water Conservation Area (WCA) 3A and the Everglades, and the non-native is there, and the native is there. I mean, look," he said, pointing to a pair of bright-pink egg clutches on neighboring stems, a few feet from stems with native apple snail eggs. Different habitat preferences were enough to explain why native apple snails were less common in areas where invasive snails were abundant. I thought of Eric Crawford saying that he sees both native and invasive snails at the pristine end of the STAs, and there were a lot more invasive snails at the inflow end where the water was dirtier.

So, snail-to-snail predation was out.

To Darby, though, the invasive snail is a red herring.

First, he said, the timing is close, but not close enough. "The native snail decline in Lake Toho happened between 2001 to 2004. The native snail population was either nonexistent or declining *before* the introduction of the invasive species." I recalled that thick mats of pickerelweed had covered the edges of Lake Toho in the early 2000s; that had been one of the problems the 2004 restoration aimed to address. Thick mats were not suitable for the native snail. Lose native snail habitat, lose the native snail, I thought.

Fortunately, Darby had already been studying the native snail's population in Lake Toho and elsewhere in the state before the invasive snail exploded on the scene. His field data exonerated the invasive snail in Lake Toho. Close in time, but not close enough.

Darby said he also considered whether climate change was to blame. Florida has certainly become warmer in the last decades. And many animals around the world have suffered from their habitats warming more than they can tolerate. But he realized that the same species of native snail is abundant in Cuba, where it is warmer than here. So increased air and water temperatures could not explain the loss of native apple snails in Florida.

If this had been a game of Clue, it wasn't the invasive snail with the candlestick in the dining room, and it wasn't climate change with the revolver in the billiard room.

We motored over to another area that he wanted to check out. This one was shallow enough for us to wade in. Darby and his field tech got into the water to look for snail egg clutches, and I slipped off the edge of the airboat to join them. Cool water bubbled into my socks and trousers.

Picking up the thread of our who-done-it, I asked, "So, what caused it? Why did the native snails decline?"

To Darby, it's not that mysterious. It's the water. Water management, to be exact. It was water management with the dagger in the lounge.

Darby stacked his hands and separated them vertically to indicate a higher level and a lower level. He smiled and said as an aside, "I'm doing my physical hand thing, which I always do in class. I'm a professor; this is what I do." If anyone could teach a lecture knee-deep in a wetland, it was Professor Darby, I thought.

He said, "The native snails are here," indicating a modest distance between his two hands. "And the invasive snails"—he widened the vertical gap between his hands, making one much higher than the other—"just have a bigger range."

I brought up Indian Prairie, the periphyton wonderland that Dr. Paul Gray had waded in with me. I said, "It seems like sometimes the water conditions can be good, and we still don't have the native snail."

Darby knew what I was talking about. He said, "Once you create a condition which is not suitable for the native snail, and it gets wiped out . . . how do you get them back? Where is the native snail coming from? And how long will it take?" This reminded me of the lag effect that Dr. Steve Beissinger had written about in the 1980s: If you get two severe drought years in a row, it can take the native snail four years or more to recover. And with water management, things can go wrong for more than just two years in a row.

Darby gave the example of 3A, the biggest part of the snail kites' Critical Habitat, where native snails were doing well through the 1990s (see figure 5.2). In fact, Rob Bennetts counted 402 snail kites roosting there in December 1994, breaking Jean Takekawa's 372-kite roosting record from 1985. Darby told me that in the 1990s, snail kites were nesting all over central and northern 3A, with more than two hundred nests in 1997 and 1998.

He said, "Then, boom! In 2001 or so they flipped the switch, and the water management plan changed. All of a sudden central and northern 3A were just dry every year. Like every year. And from a snail's perspective, you're just whacking them back. They don't have a chance."

He went on, "They were drying out central and northern 3A, and there were essentially no snails there. Not the native, and not the invasive either."

Meanwhile, the southern end of 3A got too deep. Since the native snails prefer shallow wetlands with sparse vegetation and abundant periphyton, it's important that their wetlands not get too deep to sustain that habitat. As we've seen, the invasive snails can handle much deeper water levels. I recalled that the Miccosukee Indian Tribe had lived on tree islands in 3A, and that in 2005 they sued the federal government for allowing the snail kites' prime nesting habitat to drown. The following week I would be going there to speak with a Tribal elder and see the kites' former habitat.

Darby said, "It's complicated, but the bottom line is that the native snail is right in this window, and once you get outside of that box, they're not happy. So, we managed to alter that window for the native snail there starting somewhere between about 2000 and 2003 or 2004."

He added, "The invasive snail did not show up in 3A until 2011. I know because I was monitoring the snails there. We were out there sampling for four months every single year. We saw the invasive snails in the canals down there, but that was it." Just like in Lake Toho, I realized, the timing was off in 3A. The native snail became scarce years before the invasive became established there. Water management changes do line up, though.

"Even if the native snail did come back to 3A," I said, "it seems like their reproductive cycle is so slow, it would take a long time for them to get abundant enough for kites again."

"Exactly," he said.

"Do you think captive breeding is a possible solution?" I asked.

He nodded. He had collaborated with Helen Posch and Amber Garr, at what was then called Harbor Branch Oceanographic Institute. They developed a good system for raising the native snails. In 2010, they released ten thousand native snails in Lake Toho. They chose Lake Toho because there were no native snails there by then, so any native snails they found would be the ones they released. A few months later they did the sampling to see how many survived.

Darby replied, "They found none. Zero."

He went on, "The problem was they were too small when we released them. They just got whacked. I mean, alligators eat them, turtles, fish, crayfish. . . . We just basically put out an all-you-can-eat buffet."

"Do you think it's worth trying again?" I asked. "And waiting until they're bigger before releasing them?"

He considered water conditions in 3A more promising now, and it would be worth a shot. "Because now there's more flow, and things are better. Or in Everglades National Park, which, by the way, hasn't had a snail kite nest in, like, twenty years. What if you got ten thousand or one hundred thousand snails slightly smaller than a golf ball and put them out there? They're not going to get eaten by crayfish or the small turtles. And let's just see what happens. It wouldn't cost much compared to the Everglades restoration costs, which is what, $23 billion now? I would just love to see it happen. You know, somebody should consider giving it a shot."

He sounded wistful. Conservation needs always outstrip funding. If your species is not a priority, it gets almost no money at all.

I said, "It seems like the native snail should be a priority on behalf of the snail kites, at least. They are the key prey item for an endangered species. I just don't get it."

Darby said, "The invasive snail has made the native snail sort of a moot point, in most people's minds anyway. I don't entirely blame the Fish and Wildlife Service for thinking that way. The snail kites are doing fine. They're working off the invasive snail."

I was taken aback to hear him say this. I had assumed that he, like most of the snail kite advocates I spoke with, would express distress at the snail kites' loss of the native snail and the Everglades.

His remark also got me thinking about the two hundred kites, and the idea that snail kites have lost what protection they once had, due to their numbers increasing.

I said, "Do you feel like the pressure coming from the Endangered Species Act on behalf of the snail kite is essentially gone?"

Darby said, "Not gone, but definitely people are, like, 'The kites are fine.' I think at this point you can't look at the snail kite and be worried about it. It's eating invasive snails. Is that good? I don't know. They don't distinguish between the two. They don't care. It's like saying, if McDonald's went out of business, well, hell, you've got Burger King. In 2014 we published a paper giving the snail density that is necessary if you want to keep snail kites happy so they can make babies. Other than that, snail kites don't care. They're going to eat whatever fricking snail is out there."

I struggled to put together the implications of what he was saying. "So you think that 3A is important, and the Everglades as a whole is important, and the native snail is important. But just not for the snail kites?"

He shook his head with a smile. "The kites figured it out. Thank God snail kites are so adaptive, and they fly. They figured it out. Without us.

Despite us. I can't make the 'native snail argument' anymore. Kites are finding the food they need."

I said, "Do you think the snail kites don't need us at all? That the agencies haven't helped?"

"There's a couple things that I think that the Fish and Wildlife Service and the Army Corps of Engineers did pay attention to. Water's too high. Water's too low. On the extremes."

I asked, "Do you think that they may get taken off the Endangered Species List?"

"Oh, I think there's a pretty good chance. Yeah."

"And would you support it?"

"No."

Interesting, I thought. So that was too far. "Why not?" I asked.

He pointed out that there are important parts of the kites' historical range where they can no longer be found. That would be important to address before delisting them. And the invasive snail is still a very new development. He said, "In my opinion, the data that we have on the stability of their new prey source is still unknown. Nobody is collecting data. We don't know about the invasive snail."

But Darby had had a different epiphany. And it was one I really needed to hear.

"I've been doing this for thirty years, and this came to me literally about nine months ago: The native snail is an interesting story by itself. The snail has something to say. Even if global warming and sea level rise are going to completely swamp the Everglades and turn it into a saltwater marsh, I still think there's an interesting story about the particular evolutionary history of how these apple snails made it."

I thought of the patch of bulrush he had shown me earlier that day, where the native snail seemed to be making a comeback. I had been excited for the snail kites. It suddenly hit me that I had never actually considered the native snail to have value for itself, and its own relationship to the Everglades. I saw my unconscious ranking system for the first time: the invasive snail at the bottom, then the native snail, and the snail kite above all. That's the very kind of thinking that leads to the cattail moonscape of Moonshine Bay being called a success, even while biodiversity is very low. The snail kite is not a product to be manufactured, but a living part of an ecosystem that belongs on this planet. It is part of an ensemble, not a solo player.

Darby mused, "The rest of my publications are going to talk about how we have a native species of snail. It's a good barometer of what's been going on, and maybe global climate change, or whatever you want to talk about. We should document the fact that this is a species that was highly tuned into what most people agree is one of the most beautiful ecosystems in the entire world. Billions of dollars spent, and the native snail is an indication of what it should look like, or would look like under conditions prior to human alteration."

Hearing Darby talk about it, I realized that this little snail is an encapsulation of the healthy Everglades. Where it lives on, the Everglades lives on. The fact that it persists at all is a sign of hidden strength and resilience. It is proof that the Everglades is not entirely a memento of the past but part of a vibrant present in Florida. If we can listen to this snail, we are listening to the Everglades itself.

It was tantalizing to think that the native snail had already started making a comeback. The Everglades could still regain some of its former glory.

Darby sighed. "That's what the rest of my publications are going to talk about, this native species of snail. And then I'm going to shut up, and then I'm done. And then I'm going to go fishing."

# LOSING HOME

# 25

Michael Frank showed me the chickees he had built for his family twenty years ago. Chickees are traditional Miccosukee houses, open-air square structures built on wooden stilts with palm-frond thatch roofs.

Mr. Frank, who is a Tribal elder, told me, "It was against our culture to put walls up. We could put up a mosquito net, or a tarp when a storm was coming. Then you just roll it up as soon as the storm goes, because you had to breathe clean air."

We were on a tree island called Rice Island, about the size of a suburban house lot, surrounded by miles of water three feet deep. The vast marsh was in the southern part of Water Conservation Area (WCA) 3A, just north of the Everglades National Park in South Florida (see figure 1.2). Mr. Frank told me that his Tribe calls it "Kahayale," which means "Bright Lit Place" because of the quality of light in the clear water. I imagined that the water used to be beautiful, like the rainfed wetlands of Grassy Waters Preserve and Indian Prairie. That morning, the water was opaque.

To get from one chickee to another we walked on a boardwalk that Mr. Frank had built ten years prior. Increased flooding had made it necessary; the ground below was soupy, dark mud.

The central chickee had a fire pit, with lots of burgers and sausages grilling. I had arrived via airboat as part of what the Miccosukee Tribe called their annual Fall Everglades Study, a weeklong airboat trek to visit about eighty-five of the tree islands in their homeland. At each one, they disembark and observe the condition of the island, taking measurements of water depth and quality. The Fall Study was attended by many Tribal members, as well as staff that work for the Tribe, and wildlife police. On a fleet of fourteen airboats, we buzzed from one island to the next; Rice Island was our fifth for the day. Teenagers from the Miccosukee Indian School took many of the measurements themselves. One girl wore a traditional skirt that flared out and went to just below her knees, and it featured patterns with black, yellow, and red.

Mr. Frank and I loaded up our plates and walked on the planks leading to the chickee that was for sleeping. It had a large, broad, table-like platform, and the thatch reached down almost to that level. We sat there enjoying the shade and the cross-breeze. He said that when there was a storm, people could sleep there, protected from the wind and rain by the low-reaching roof. He pointed up to the many support beams in the corners of the roof and said that it was hurricane proof. It must have been; the Fall Study had been postponed for a week due to a hurricane blowing through, and it was none the worse for wear.

Mr. Frank told me that he had built the chickees here so that his nephews and nieces could experience how their grandparents had lived and learn the traditional ways. He knows which tree island his grandmother was born on at the turn of the last century, before the big east–west highways came along. US Highway 41, often called "Tamiami Trail," was built in 1928. It cut right through the southern end of their homeland, and stopped up water flow. By the time Mr. Frank was born in 1957, the Everglades had become sliced and diced with levees and canals. In 1962, the Miccosukee Indian Tribe's homeland was enclosed by levees to become part of 3A.

The name of the tree island that Michael Frank was born on translates to High Island or High Ground. When he was growing up, it was only wet when it rained hard, and the water drained off soon after. Back then, he told me, snail kites were "all over the place." So were the native apple snails. Members of the Tribe would catch the pretty brown snail shells, clean them out, and sell them as souvenirs. His family had everything they needed. They grew crops like corn. They went frogging, fishing, and they hunted deer, alligators, wild hogs, and birds with rifles, spears, and bows and arrows. They only came out of the marsh to get salt, sugar, flour, or other staples at a trading post. In the winter, migratory birds would descend like a white cloud to roost on their tree islands.

But they don't do that anymore.

In 1968, 3A was cut through by a new east-west highway I-75, aka "Alligator Alley." The roads blocked the flow of water, causing it to pool deeply. The sheet-flow wetland had turned into a series of reservoirs. Gradually

FIGURE 25.1. Miccosukee Tribal Elder Michael Frank on a tree island in 3A where his uncle used to grow corn in the 1970s. He is standing in several inches of water on an island that in his youth would rarely have been wet. Now it's flooded for several weeks every year. Behind him, sugarcane is growing. Photo by Hilary Flower.

the tree islands started to drown, even High Ground. Although water is supposed to be cleaned in the Stormwater Treatment Areas (STAs) before entering the WCAs, several canals and pump stations dump untreated farm runoff directly into 3A.

Mr. Frank says the water became too polluted for him to drink, and too deep for wading birds. He stopped hunting and fishing in the 1970s, because his Tribe only fished for sustenance, and all the birds and fish had become toxic from mercury.

Mr. Frank's family watched as other families gradually packed up and left the tree islands for neighborhoods with roads and mailboxes. He told me, "Our family was the last one to come out of the Everglades, in the late

1960s. The island was going underwater. Worse and worse every year. So," he said, "we came out." He shrugged sadly. "Who's going to live in mud?"

His family continued to spend as much time on the tree islands as they could, but his life had changed forever. It's a loss he does not have words for. He waved a hand and dropped it. "They have made it so we can't worship our God."

"What do you mean?" I asked.

"Our creator taught us how to survive in the Everglades, how to plant, how to do everything. And now we can't do these things." Mr. Frank said, "3A is our land. Our home. And we never gave it away."

The Tribe leases 3A based on a 1982 agreement in which the federal government pledged to do three things: (1) to preserve the leased land "in its natural state for the use and enjoyment of the Miccosukee Tribe;" (2) to preserve "fresh water aquatic life, wildlife, and their habitat;" and (3) to "assure proper management of water resources."

Mr. Frank considers these all to be promises broken. He meets often with government agencies and urges them to clean the water and let it flow through the Everglades. "The canals are like veins," he said. "If the blood is poisonous, your heart is pumping poison to the different organs of your body."

The Tribe's data from Fall Studies going back to 1982 showed that water levels are much higher now than in previous decades.

Mr. Frank said, "Our islands are all underwater. That's why the trees are dying. That's why there's no rabbits, raccoons, deer, anything. It's just a skeleton." He hasn't seen a snail kite in more than a decade.

I learned more about the missing mammals from Marcel Bozas, the director of the Tribe's Fish and Wildlife Department. He wrote his dissertation at Florida International University on how mammals used the tree islands in the Everglades. On the drowning tree islands, he only saw two mammals: Florida black bear and rats.

Indeed, we'd seen bear scat at every single tree island we'd come ashore on that day. One on island, Bozas and one of the high school boys squatted down to poke at a dark-black Frisbee-sized disk, surmising that the bear had been eating turtle eggs. Nearby, Bozas showed me lots of holes full of water and floating white blobs that he told me were the remains of turtle eggs. The bears can smell where they are buried, and they dig them up. On one dock we saw what looked to me like someone had poured pumpkin seeds in three flat piles. Bozas said they were evidence that a black bear had eaten its fill of pond apples from that island. They can't resist the fruit, but it gives them the runs.

Bozas said that in the more northerly part of 3A, where it is much drier, opossums, raccoons, bobcats, and panthers heavily use the tree islands. One factor was by far the biggest predictor of mammal use in the tree islands of his study: the water levels. Deep water is great for gators, but it keeps mammals out. The too-deep water was causing a drop in biodiversity.

The Miccosukee Tribe saw that the endangered snail kite had been pushed out of the 3A. In 2005, the Tribe sued the federal government on behalf of the snail kites, in the hope that the Endangered Species Act (ESA) would carry more weight than the 1982 agreement.

This was not the first time that water management and the ESA had clashed in 3A. On February 13, 1990, the US Fish and Wildlife Service (USFWS, or "the Service") hauled off and used the strongest word in their arsenal: jeopardy. They issued a formal biological opinion stating that the planned change in water management in 3A would jeopardize the continued existence of snail kites. It was crucial not to further degrade snail kite habitat in 3A because it was the most important kite nesting stronghold since monitoring had begun in the 1960s, and by far the largest part of their Critical Habitat as designated in 1977 (see figure 5.2). Unfortunately, the region in 3A that was their nesting stronghold was the same area that would become the deepest.

That jeopardy opinion, however, did not stop the Army Corps of Engineers from going forward with their plans. The Service simply required the Corps to provide funding for a snail kite monitoring program, in collaboration with the US Geological Survey. Dr. Wiley Kitchens was tapped to be its director. If the purpose was to make sure the snail kite population did not suffer negative effects, there was ample evidence of that by 2005. The 2005 report from Kitchens's team highlighted "alarming trends" in snail kite populations and cited water management in the WCAs as a principal cause: It was the too-deep water, as well as too-rapid decreases while snail kites were nesting. Only thirty snail kites fledged in 2005, and not a single one came from the WCAs.

The Miccosukee Tribe's 2005 lawsuit aimed to hold the US government accountable for the harm done to the snail kites. In response, the USFWS wrote a biological opinion that echoed the Miccosukee's points about how the snail kites were being harmed by the water management plan for 3A (known as the Interim Operation Plan, or IOP). The document states: "Continued IOP operations are expected to result in continued habitat degradation within 3A, which has been one of the most significant areas of kite habitat within the past 30 years. In addition, IOP operations are expected to result in reduced nest success of kites within 3A, reduced foraging habitat suitability, and reduced abundance of the kite's primary prey. These impacts are expected to limit population growth in 3A and possibly cause further reductions in the overall kite population."

Then they pulled the rug out: "However, because snail kites are long-lived, have high rates of adult survival, and continue to successfully nest in other portions of their range in southern Florida, these impacts are not anticipated to appreciably reduce the likelihood of survival and recovery of the species in the wild. Degradation of designated Critical Habitat within 3A is expected to continue under IOP, but this is reversible with improved hydrologic conditions. No permanent loss of Critical Habitat is expected."

I was dismayed to see them using the kites' resilience and adaptability as an argument *against* protecting them, even in their Critical Habitat, even when their population numbers were in free fall.

The biological opinion also asserted that alleviating the too-deep water would harm an even more imperiled species, the Cape Sable seaside sparrow. I spoke with Jaclyn Lopez, who had been the head of the Center for Biological Diversity in Florida at that time. She said that the sparrow continues to be used as an excuse for keeping 3A deep, and she shared with me a letter she wrote in 2020 to the South Florida Water Management District (SFWMD). She pointed out that one of the main goals of restoration was to provide *more* water to flow from 3A into the Everglades National Park, since it's much drier now than in the predrainage condition. She asserted that the sparrow would be a beneficiary of this, not an obstacle.

If even the Florida director of the Center for Biological Diversity thought the sparrow was being used as a scapegoat, something else must be at play, but I couldn't find it in the biological opinion. I had reached out to Ed Ornstein, deputy general counsel for the Miccosukee Tribe. "What am I missing?" I asked Ornstein. "It doesn't make any sense."

"When things don't add up," he replied, "it means you're missing a piece of information."

I knew that there was some debate about how deep the water had been in that part of the Everglades before drainage, and whether land use on tree islands may have affected the ground elevation and contributed to drowning. But the biological opinion did not raise these points; something else was at play.

As I dug into it with Dr. Paul Gray of Audubon Florida, water supply seemed the most likely—unspoken—reason to keep 3A deep. The WCAs were constructed as reservoirs, after all. During droughts, water in 3A can be directed via canals to fulfill municipal, urban, and agricultural needs. Deep water in 3A also replenishes the very porous aquifer and keeps the water table high. This in turn keeps water in canals miles away, and helps push back against saltwater intrusion near the coast. Allowing 3A to be deep also means it can take more floodwaters from upstream, enhancing flood control. Another factor is that one of the water control structures needed for letting the water out of 3A was constructed at too high of an elevation; it would require a costly modification. Finally, there's a concern that if they let water flow out of 3A, there would not be enough clean water upstream to replenish it, and 3A could become chronically dry. In a decade or so this problem may be somewhat alleviated by a new Everglades Agricultural Area (EAA) reservoir and an adjacent STA. But only somewhat.

Paul Gray said that because water is being held back in 3A, the Everglades National Park to the south is way too dry. He said, "Many blame the sparrows and the control structure for the crime, but outflow is but one part of the equation, and ignores water inputs and management decisions to hold it as high as possible. Drowning tree islands has been a crime against the Earth."

Ed Ornstein said that things might be shifting. In November 2021 at the Tribal Nations Summit, the Biden administration announced that all federal agencies would be required to incorporate Indigenous Knowledge into their research and policymaking. Ornstein told me that he felt a new optimism that the Tribe's concerns about the too-deep conditions in 3A could be addressed if Indigenous Knowledge is respected. Of course, the following administration could easily undo all that progress. Although many of former President Joe Biden's executive orders and memoranda have been rescinded, this one has been spared, at least for now.

Ornstein said that the Miccosukee Tribe had cared for the ecosystem for generations. The Miccosukee Tribe descended in part from Indigenous people who migrated to southern Florida to flee genocide. In the Seminole Wars of 1817–58, US soldiers tried and failed to root them out of the Everglades. Several sources from that time indicate that they may have allied with and intermarried with Indians from South Florida, like the Calusa Indians, who lived in the Everglades for thousands of years. Indigenous peoples have a longer relationship with this land than the Everglades itself. Native people arrived 13,000 years ago, but the Everglades didn't begin forming until 5,000 years ago—making them witnesses and participants in its emergence. The oldest tree islands are 3,000 years old, and most are younger than 2,500 years old. Research has shown that Indigenous people helped grow them. They sought out patches of dry ground when they came into the interior of the wetland to hunt and forage. They brought coastal food with them like shellfish. In so doing, they contributed enough fertilizer and substrate like pottery and shells to help build the tree islands we have today. It's a remarkable example of humans slowly shaping iconic features of the landscape in the past.

Now, because the soil on tree islands in that area has become saturated for much of the year, the trees have become unstable and more likely to topple in storms. Mr. Frank said, "When a tree dies, that's the foundation of the tree islands. It gets washed away every year."

We came on a fallen tree on one of the tree islands. Bozas said that anytime he sees a fallen tree on a tree island, he has to go see what he can find among the roots. Since the trunk was lying on the ground, the underside of the root structure made a wall of roots and dirt. He pointed out white objects visible amid the soil. "That's a pottery shard," he said, pointing to an angular fragment. "So is that." He looked around, his fingers touching the soil. "That bone is from a gator." It was an angular fragment like the pottery shard, but it had a pattern of divots in it.

"And that bone," he said, pointing to a tubular shape about three inches long, "that is probably raccoon."

He dug a little with his fingers. "Oh," he said. "Look at this, it's a seashell." He handed it to me.

It was a fragment of a scallop. I had seen mussels in freshwater wetlands but never scallops. I said, "So, people brought this scallop here from the coast, many miles away, and ate it here?"

He nodded. I had read about this happening, but it was different holding an actual scallop that someone brought to this spot hundreds or even thousands of years ago.

Bozas said, "When you dig down in the soil on all of these tree islands, it's just like this: shells, pottery, and bones, layer after layer."

The Everglades is important for its history, for what it has been and can still be for the humans and wildlife that have grown alongside it, serving it, and being served by it. The Miccosukee Tribe doesn't just *want* it to be healthy, they *need* it to be healthy.

Michael Frank said, "Living off the land, that's how we survived the soldiers, and that's how our creator told us that we were going to survive. Stay here in the Everglades, live off the land, drink the water, eat the birds, the fish, and the plants, and whatever is growing in the Everglades. But be Indigenous, be a part of it. Our creator will look down on us and see we're still in the Everglades. We've still got the chickees, living in our traditional cultural ways. Our maker told us to stay here, and his hand is upon us. But if we leave the Everglades, and we assimilate to the outside world, then, unfortunately, our way of life and the Everglades will be destroyed."

As our airboat started heading back to the docks where we had started, I watched as the marsh sped toward us, and we passed tree island after tree island. I realized I had gained a much clearer view of the Everglades' past and its present, and a new, very tentative, hope for its future.

Bozas said that the Tribe has an aquaculture lab, and they are raising native apple snails. About one thousand per year. They hope to repopulate the wetland with apple snails. It's not too late. If the ecosystem comes back to life, the snail kites will return.

For decades, the Service has justified the degradation of the Everglades based on it being "reversible." It seems long overdue to make good on that implicit promise. Reverse the degradation before it's too late.

For a lot of people, the Everglades is a relic, obsolete, a museum exhibit, a curiosity, at best a wonder to be gazed at, at worst a long stretch of road without gas stations.

But for some, the Everglades is more than all of this.

It's a home.

# NEW FRONTIERS 26

When I tried to imagine the snail kites' future, I kept thinking about something FWC's snail kite coordinator Tyler Beck had said out at Moonshine Bay: Snail kite population numbers seemed to have plateaued at about three thousand since 2017. It seemed like the snail kites were at a crossroads. What was next? Given their wild ups and downs, anything seemed possible, from extinction to full recovery, or simply bouncing very close to those two extremes indefinitely.

Going forward, it seemed that snail kites could expect little active help unless their population were to plunge to the very brink of extinction. Tyler Beck had recently told me that the hydrilla policy had changed at Lake Toho: invasive plant managers have found that it's really not possible to take a tolerant approach to hydrilla after all. On a given lake, they must either give it free rein or eliminate it, and be vigilant about keeping levels extremely low. Perhaps it is just as well that in recent years snail kites have not been nesting as much in Lake Toho as they used to.

When it comes to growing the snail kite population by expanding available habitat, the Everglades was the obvious wild card. The snail kites' future would brighten if the Water Conservation Areas (WCAs) became healthy enough to support the native snail and, with it, the snail kite's return. Grassy Waters Preserve burns like a candle, proof that the vital relationship between the snail kite and the native snail still survives in places. Would the thriving population of native snails near Ox Island in Lake Kissimmee grow and spread? And when, if ever, would the water impounded in 3A finally flow into the Everglades National Park as intended? Though the kites have vanished from the Everglades, I still scan the skies for them whenever I visit, hoping they will come back.

Of course, the invasive snail was another big wild card. It seemed like we might be at the end of a statewide twenty-year boom-and-bust arc. The snail's biggest booms and the kites' biggest nesting events seemed to have been early in their invasion. The invasive snail is now prevalent in much of central and northern Florida, and their normal "background" population alone is not enough to inspire much snail kite nesting.

There were two other wild cards that seemed likely to shape the snail kites' future: climate change, and a shift north beyond Florida. I pursued these with four experts, each of whom held a piece of the puzzle.

Climate change kept nagging at me since I am a climate scientist. As snail expert Dr. Phil Darby had noted, warming alone is not likely to make much difference, since the snail kite and the native snail also live in the warmer climate of Cuba. But warmer temperatures suck moisture form water, soil, and plants, making droughts more severe. Given the tendency of drought to cut snail kite populations in half, it struck me that changes in evaporation and rainfall in the coming decades will be extremely impactful for snail kites. It's still uncertain whether Florida is likely to get more rain or less rain.

I had heard that a researcher named Dr. Kathryn Smith had done some climate scenarios modeling for snail kites, and I was excited to talk to her about it. On my screen Smith had wavy, light-brown hair and a friendly, open expression. She was the lead producer of Species Status Assessments (SSAs) at Texas A&M Natural Resources Institute, under contract with the US Fish and Wildlife Service (USFWS). When we spoke, she had recently completed the SSA for the snail kite, although it hadn't been released yet. She had done previous assessments on crayfish, lizards, fish, and plants. But she said the snail kites had been her favorite one. Before talking climate change, I had to find out why.

"The snail kite is such a complex species," Smith said. "I enjoyed absorbing everything about them I could." She went on, "How many species just go wherever they want every year? They're like, 'This isn't any good, even though I was here last year. I'll fly hundreds of miles and nest at another place.'" She laughed. "It's pretty incredible." I couldn't argue with that.

I said, "I heard you made some forecasts about how snail kites might fare with future climate change."

Smith was ready to jump in. "Yes. We looked at historical rainfall for years when we knew nesting outcomes. To get climate scenarios, we teamed up with the US Geological Survey. They said that it could be a lot more rainfall in the future. Or a lot less rainfall. So, we just took the extremes: What's the worst-case scenario, and the best-case scenario? And then probably reality is somewhere in between. But you want to know your bookends."

She went on, "And it's not just how much rainfall, it's how variable it is year to year. With climate change, you're going to have more extreme weather. You're going to get some drought years in a row. You're going to get some really wet years in a row."

She counted off on her fingers, saying, "So, we did three scenarios: One was 'highly variable,' one was 'less rainfall and more drought,' and the third was basically 'no change.' For the 'no change' scenario we used exactly what happened in the last twenty years."

Scenarios modeling is a way of exploring different possible directions a system can go. I had done some scenarios modeling of mangrove distribution and soil for the Everglades. But the model I used could not say anything about snail kites.

I said, "And what did you find?"

"So, unfortunately," she began, "what we found was in all those scenarios, snail kites don't do well."

She shook her head with a sympathetic expression. I realized that giving bad news was one of the hazards of doing species assessments.

She said, "It was just not good. They don't do well with extremes."

"Now," she said, with a raised finger, "that's not taking into account a possible shift north. We don't understand that yet, and that might be a solution. I mean, they adapt the way species do, and there's no way to predict it."

After our conversation, Smith emailed me her climate analyses. For all the climate scenarios she had mentioned, I could see that the snail kite population shrank.

But there was one more scenario, and it offered a more optimistic outcome. It was a scenario in which the effects of climate change were buffered by extraordinary conservation efforts on behalf of snail kites. It portrayed a world in which Everglades restoration was complete, and there was snail kite–friendly water management. In that scenario, the snail kite population increased, although snail kite resiliency was still rated "low." Smith had probably not mentioned it because it was so unrealistic.

It seemed unfair to expect wildlife to adapt and survive without us taking an active role in addressing climate change and other human impacts. But thus far, I didn't have a lot of evidence that humanity was going to mobilize on those fronts any time soon, especially in the US.

Smith's point about a shift north was intriguing. Lots of species are shifting their ranges poleward as temperatures rise. Rough-legged hawks and golden eagles have been moving north by five miles per year since the 1970s. I could imagine that if droughts were to become more severe in Florida, snail kites might well explore their options to the north.

The mere fact that the snail kites have already followed the invasive snail into the northern reaches of Florida makes a northward shift even more plausible. In the last decade, the invasive snail has expanded beyond Florida into Georgia, Louisiana, Mississippi, Alabama, Texas, Arizona, North Carolina, and South Carolina.

So far, only a few juvenile snail kites have ventured north from Florida, and they don't seem to be lured by snails. On eBird, the app for bird sightings, I found several reports in the last few years of juvenile snail kites seen for a few days at a time along creeks or in crayfish farms in Georgia, Texas, and South Carolina. None of them were described as eating snails, only crayfish. And it doesn't seem to end well. Kayakers found a weak and emaciated juvenile snail kite along a creek bank in Tennessee and took it to a rehab center.

Still, limpkins may be a bellwether for future snail kite movements. This wading bird is the other predator that specializes in apple snails, although not to the extreme that snail kites do. The limpkin had been listed as endangered in Florida; I remember the thrill of spotting a rare limpkin in the Everglades in 2010. Since then, the invasive snail has made them commonplace along retention ponds and wetlands throughout the state. The USFWS removed the limpkin from the endangered species list in 2017. By that time, they had expanded their range north. The first eBird report of limpkins outside Florida came from Georgia in 2012. In the next five years, limpkins showed up across several states along the Gulf coast. When they started breeding outside of Florida, that was a noteworthy development. Limpkins successfully reared young in southern Georgia starting in 2016.

Dr. Jacoby Carter was the first to document the invasive snail's arrival in Louisiana in 2008, followed by nesting limpkins a decade later. So, he was the next expert I wanted to speak with about the snail kites' future. He spoke to me by video from Lafayette, Louisiana, where he has retired from the US Geological Survey (USGS). He emphasized that he was not speaking on behalf of the agency but sharing his own observations and thoughts.

When I asked him if he thought the snail kites could follow the invasive snails and limpkins north, he didn't hesitate: "Yes."

He said, "I can't see any reason snail kites can't expand their breeding range. I've been saying this for about ten years. I saw it coming. It's just basic biology. If a species' range is limited by food availability, and a new food resource over a greater range is available, then it seems like their range would expand. Unless something else is restricting their range."

Carter went through a checklist of potential factors that could limit snail kites' range expansion. Predation? He said he couldn't rule out the possibility that great horned owls could eat more adult snail kites, or that more animals could prey on eggs and young in nests. But he doubted it.

He considered nesting habitat. If there were not enough short trees or patches of cattail near kite foraging areas, or if water below the nest dried up too quickly, these factors could limit nesting success.

The more he thought about it, the more water clarity for snail-spotting seemed the most likely barrier. He mused, "If you can't see it, you can't catch it. That would leave out Louisiana and parts of Mississippi, although it may just be a matter of time before they find spots with clear enough water." He said that there may be water sufficiently transparent for snail hunting in the ponds and streams of Alabama, and also in Texas, if snail kites could get there without much foraging in between.

In contrast, limpkins were not limited by water clarity, because they don't rely on seeing their prey. And many eBird reports of limpkins beyond Florida indicated that they were consuming freshwater mussels rather than snails. But Carter assured me that the limpkins breeding in Louisiana were eating snails. It also occurred to me that limpkins can make do with lower densities of snails compared to snail kites, which has allowed them to become common across the state, including on my college campus. Yet, for kites to breed farther north, invasive snail booms may be required.

Dr. Paul Gray of Audubon Florida is the third expert I wanted to speak with on the future of the snail kites. He's the person who connected me with Jacoby Carter, so I shared Carter's response and asked for his own.

Gray thought seasonality would pose the biggest barrier to the kites' expansion northward. Cold temperatures cause snails to become inactive, and they stop coming to the surface to breathe, where snail kites could spot them. If it gets too cold for too long, snails can die, especially young ones.

Gray said that at the very least, the kites would have to migrate south for the winter. He said, "They'd have to learn to be seasonal migratory birds."

I had dug up some support for this possibility. A 1980 report documented snail kites in northern Florida shifting south during particularly severe winters. Also, two snail kites of the South American subspecies migrated seasonally over two thousand miles from the southeastern corner of Brazil up to the mouth of the Amazon River. Still, Gray was skeptical. He said, "Maybe they could do that, but for a tropical species, who knows?"

Another consideration is that many of the more northern wetlands used by limpkins and invasive snails are popular multi-use areas. The ESA applies wherever snail kites are found. But the new areas would have no experience or protocols for protecting snail kites, so protection might involve growing pains like Lake Toho's Hydrilla War. Plus, given that snail kites have found little protection even in their Critical Habitats, I would expect even less enthusiasm in states outside of their historical range.

An added dynamic is that range-shifting can cause a species to become delisted and lose ESA protection entirely. The Service has recently removed the wood stork from the endangered species list in part because it is feeding in the rice paddies in South Carolina, even though it no longer nests in its historic range in Southwest Florida.

Looking to the snail kites' future, I felt confident that there would be some big surprises ahead, one way or another. I realized that, at the heart of my quest, I was seeking reassurance that our own species would be okay. Although our species dominates the natural world, we rely on it, too. We are vulnerable in ways that are often overlooked. The snail kites' agile adaptability far outstrips our own. We could learn a lot from them.

There was one more person I wanted to talk to about the question of the kites' future: Dr. Steve Beissinger. After all, he had been one of the first to discover just how adaptable snail kites can be.

In our next video call, I asked him if he was hopeful for the snail kites' future. He immediately smiled and reflected on his introduction to them, up a tree with Noel Snyder in 1979: "Honestly, we spent a lot of time sitting on that platform watching them. And they'd always surprise you." He said that they might spend all morning going to one place for snails, only to abruptly change to another spot. He said, "What's that about? Why are they doing that?" He laughed. "There were always different things like that. They would just surprise you."

"No one would have expected the kites to switch to a different snail and a different habitat," Beissinger continued. "So, if they can do that," he lifted his hands, palm up, "and it can work out, then that's it, isn't it?" He shook his head and smiled. "Should I feel bad that they're not feeding on the native snail? I don't. It's the Anthropocene now."

I had to laugh. The snail kite really was a poster child for the Anthropocene, a term coined to capture the observation that humans have impacted the Earth to such a degree that we have launched a new geologic era. I was also interested to hear Beissinger echo snail expert Dr. Phil Darby's equanimity about the invasive snail taking the place of the native snail for the kites, at least for now. I saw their point. Humans have changed the rules as to how the planet works, and the snail kites are surfing on the cutting edge.

With raised eyebrows, Beissinger said, "It's hard to understand exactly what baselines mean in a world where they're all changing. The position of species is changing, with both the fluctuations of non-natives coming in, and expansions of ranges due to climate change and things like that. So, it's . . ." He shrugged with a bemused expression. "It's a different world."

"And," I said, "they're going bravely into it."

"Yes," he said, nodding emphatically. "They are."

# TO CATCH A KITE 27

I often thought back to the days in Loxahatchee Slough Natural Area in Palm Beach County (see figure 1.2) with Gina Kent from the Avian Research and Conservation Institute (ARCI). I gained a vivid understanding of snail kites on that trip, watching their every move, with Kent there to interpret for me. From the beginning of this project, I held a secret hope that I might get to hold a snail kite in my hands. I didn't know why, or what difference I thought that would make. Somehow it felt like an important part of my quest for hope and for understanding the plight of this endangered raptor. I eagerly awaited an email from Kent announcing her next trip.

Finally, the day had come. In the parking lot, I met up with Kent and Jenny Bouchenot, a graduate student from the University of Central Florida.

Kent told us that there were a lot fewer kites in the area than during our last attempt, due to Hurricane Milton hitting the area. The communal roost had only fifty-four snail kites, down from 146.

We drove around together in Kent's field truck, on the lookout for silhouettes of birds in trees and a flash of white on the tail of a flying bird. We checked some of the most promising areas from the last trip, with no luck.

We tried the lone cypress tree where one had been perched last time. We were delighted to find a female kite perched there. Kent stood on the levee scoping it out, to see if there were other perches she used, and to make sure she was at least two years old. The kite flew away for a while, and we settled into camp chairs to see if she would come back.

As we chatted, I asked Kent, "What specific things do you think humans have done that have meaningfully helped kites? What do you think we *should* do to help them?"

She thought about it and replied, "Protect wetlands."

I waited for her to say more.

She looked around thoughtfully, as if reading the answer in the cypress trees and lily pads in the wetland before us. Then she nodded slowly. "Yes. Protect wetlands. That's the most important thing we can do."

The female kite came back. I resolved to take time later to hold up Kent's two-word answer to the varied snail kite experiences I'd had throughout this project.

We watched our female snail kite perch on a smaller cypress tree for a few minutes. Abruptly, she flew off to grab a snail from the marsh and then returned to the lone cypress. I trained my binoculars on her as she held the snail in her bill. She tried to position it properly with her feet so that she would not lose her balance. It took a few tries. Then she tilted her head and began digging into the snail with her bill.

Bouchenot and Kent talked about how one of the reasons for trapping and tagging snail kites was to collect feather samples for methylmercury analyses. This naturally occurring element is toxic, and human sources such as trash incineration can raise levels in the environment. Previous work by ARCI's Dr. Ken Meyer and Kent showed that methylmercury concentrations in snail kite feathers increased as one headed southward in Florida, with the greatest levels in the Everglades. The southernmost concentrations (up to 4.5 and 4.8 ppm) had the potential to kill kites in the egg stage. Meyer wondered if mercury exposure could be a factor in the snail kite's severe declines in the Water Conservation Areas and the Everglades National Park. ARCI's next step is to learn whether snail kite eggs in the Everglades have lethal loads of methylmercury, and how egg concentrations relate to maternal loads. Bouchenot would be doing the lab analyses as part of her graduate work. Wow, I thought. That could be an additional cause for the snail kites leaving the Everglades. I was glad that ARCI was on the case.

I had learned a bit about the purpose of the GPS transmitters from the most recent Snail Kite Coordinating Committee meeting. Meyer gave a talk in which he showed a map with red dots corresponding to places their tracked snail kites had been, going back to 2007. It looked like someone took a paintbrush and splashed red up and down the length of the Florida peninsula. The red seemed especially thick in the Everglades, south of Lake Okeechobee. That made sense to me, because I'd seen them in the Everglades myself through 2013. Meyer said that in the early years their tracked snail kites spent about 60 percent of their time south of Lake Okeechobee, and used their whole range.

But ARCI's tracked kites rarely go south of Lake Okeechobee anymore. Meyer is hoping that the US Fish and Wildlife Service will take heed and expand the snail kites' Critical Habitat northward. They need wetlands beyond their Critical Habitat now more than ever (see figure 5.2). More importantly, Meyer wants them to address the decline and near loss of kites from the Everglades. He told me he feels discouraged at how few management efforts have actually been expended on the snail kites to date. And he is frustrated at the many conflicting and corrupting influences that affect how agencies respond. "The lack of meaningful response is disgraceful, at the least," he said. "And very few people seem to notice or care, professional or lay public."

Kent tapped a bit on her phone and then showed me the screen. It was another map with colored dots. The dots showed the movement of the two kites that she had trapped earlier that year in the same area we were currently in. One had gone up to the St. Johns River, an old snail kite haunt almost two hundred miles north of where we were.

ARCI's tracking data have been important for protecting the snail kites. The state has been trying to extend a state road all along the western edge of Grassy Waters Preserve. The City of West Palm Beach, which owns the preserve, argued that the big road could impact water quality. Having a major north–south artery so close to snail kite nests could also be disruptive. I hated to think of anything spoiling the unofficial snail kite sanctuary of Grassy Waters. The state had downplayed the preserve's importance to the snail kites, pointing to other wetlands as alternative habitat (or "mitigation wetlands") for snail kites to use instead. That's where ARCI's data came into play. As part of the city's lawsuit to stop the road, Meyer testified that the GPS trackers showed snail kites spent a lot of time in Grassy Waters and Loxahatchee Slough Natural Area, while almost entirely ignoring the alternative "mitigation" wetlands. Meyer said that one young kite had settled into a corner of one mitigation area briefly, and that was about it. By putting GPS trackers on snail kites, ARCI had given these endangered raptors a voice. And their message was clear: Grassy Waters matters to snail kites. It needs protection. The "mitigation" wetlands were not valid substitutes.

The final judgment had not yet been handed down. When I had talked to Meyer about the lawsuit, I could see the pain in his face. Meyer is a fighter.

He loses sleep at night, always trying to figure out the best way to solve a problem. Together Meyer and Kent have been a force to be reckoned with for twenty-five years.

I asked Kent, "Do you feel hopeful for the future of snail kites, and other wildlife?"

She considered: "Yes, I have to. When I'm out here, I feel hopeful. There's a lot of bad news out there, and I don't make a point of keeping up with all of the different kinds of environmental problems. It's too much. I mean, I follow it enough to be aware, and I vote, but there's only so much I can do about those things. There *are* things I can do for snail kites and the other birds that Ken and I work on. So, I just focus on doing that."

I admired her quiet resolution. She does not get tripped up by the messiness of it all. She does not crumple in despair over the enormity of it all. She knows what she can do, and she does it with all of her heart.

A few minutes later, Kent brought her binoculars down and grinned at Bouchenot and me. She said, "I think this is good. When she flies off, let's put out some traps."

We brought the kayaks down from the top of the truck. Bouchenot put her stuff in one, and Kent the other. Kent stacked the traps on the back of her kayak.

When the snail kite left her perch, Kent and Bouchenot eased the two kayaks down the levee, slid them into the water, and paddled out. They set the traps at the lone cypress perch and the smaller cypress. They paddled off to watch at a distance, and I watched the traps from the levee in a camping chair.

I settled in for what could be several hours of waiting, watching, and hoping. I thought about Kent's deceptively simple "protect wetlands." I remembered Jean Takekawa saying, "There's still a lot left to protect." I was interested to note that this matched the "Secret of the Everglades" that Sam the Snail Kite discovered in Vera's children's book: Protect the water, and protect natural areas.

I searched my mind for greater complexity. Surely I had found more nuance, more rigorous detail in what it would take to really help the kites, more confusing strands to grapple with, more agency- or policy-based solutions. After all of my digging, I discovered that the root problem was the banal and seemingly intractable problem of government agencies being subject to moneyed interests. And the inherent tension of a wetland peninsula being forced to behave as dry land, with the residual wetlands being pressed into service as reservoirs.

The times that humans helped snail kites were all examples of protecting wetlands. The heartbreaking failures to protect snail kites were all failures to protect wetlands. Regardless of outcome, the fight is still worth fighting, for agency people looking for compromises, for the Miccosukee Tribe in their Bright Lit Place, for the City of West Palm Beach and ARCI trying to protect Grassy Waters, for Dr. Paul Gray of Audubon Florida refusing to accept the degradation of Moonshine Bay, for Tyler Beck of FWC and the many people at government agencies finding compromises on behalf of snail kites, and for all people who speak up for wildlife. Even crushing defeat can lead to unexpected wins down the line. The Hydrilla War of 2008 ended in failed nests that year, but it led to the creation of the snail kite coordinator position at FWC, and a bumper crop of snail kite fledglings at Lake Toho in 2011.

Protect wetlands: I could not find anything in the long saga of snail kite conservation that fell outside of that umbrella. Protecting wetlands can become complicated and at times maddening, of course, when dealing with human systems. But protecting wetlands is itself a specific and concrete mandate, a North Star. As far as we let those two words guide us, we bolster biodiversity and protect endangered species—not as single isolated species but as part of a functioning and flourishing ecosystem.

The exquisite blue and green wetland I was gazing out at from the shade of the field truck was a shining example, and a surprising one. When Jean Takekawa left Florida in the 1980s, this was a wasteland of abandoned pasture and ditched agricultural fields, running amuck with invasive plants. The spot where I was sitting had been a monoculture of an invasive tree called melaleuca. Looking out at the clumps of cypress trees surrounded by endless blue water with bright-green lily pads, I tried to imagine it as a dense stand of the tall, skinny melaleuca trees.

Given how many wetlands were lost after Takekawa left Palm Beach County, it gave me deep satisfaction to learn that one major loss had been radically reversed. Within a decade after Takekawa's departure, Palm Beach County began acquiring and rehabilitating the 13,025 acres that now make up Loxahatchee Slough Natural Area. Melissa Tolbert manages it for Palm Beach County's Department of Environmental Resources Management. Tolbert had shared this success story with me by phone from her office at the county. She said that in twenty years, she has been fortunate to witness the full transformation. As the habitat restoration progressed, snail kites and other wildlife were quick to move in. The Loxahatchee Slough Natural Area was the biggest snail kite nesting spot in 2020.

I have started sharing the story of the Loxahatchee Slough Natural Area with my students, as a reminder that environmental degradation is not on a single track toward worse and worse. Every now and then, humans turn bleak into beautiful. Humans, too, can be surprising.

Watching the snail kite on the little cypress tree a hundred yards in front of me, I thought about how she was one of the many beneficiaries of these efforts. I lifted my binoculars to get a closer look at her. She stared down at the trap for a long time. I thought about how back in 2008, Sal had pointed out my very first snail kite on a little tree just like this. I wished that Sal could be there to see this snail kite, so content in this little recovered spot of the Everglades.

The snail kite left her perch. I sat up. She hovered over the trap for the first time. I was not sure if Kent could see, so I texted her "hovered." It was the first time that day, so it felt important. A good sign.

The kite continued to hover . . . dropped a bit . . . flapped to go back up, but she did not. She was still at the trap.

Could it be?

Yes!

We caught a kite!

# THE HEARTBEAT OF A KITE

# 28

I called out to Gina Kent, who was already paddling vigorously toward the trap. My heart racing, I backed the pickup truck to be at the closest spot to where Kent would arrive at the levee. I lowered the back of the truck, making a little work surface. Within ten minutes Kent had secured the kite in a protective tube, paddled over, and climbed up onto the levee. Bouchenot paddled around to collect the other traps, and joined us by the truck.

To avoid further stressing the kite, Kent and Bouchenot started speaking only in whispers. Kent laid the tube down on the back of the truck and motioned for me to put a hand on the tube to steady it. She gently pulled one bright-orange foot out at a time to take measurements.

I squatted down, and with my other hand I gently touched the dinosaur-like pattern on the cool skin of her feet. The black talons had a polished luster and an unexpected elegance. Bouchenot whispered that when most raptors land on your arm, they can break the skin, and she appreciated that was not the case for snail kites.

Kent put a silver band on the kite's left leg, a federal band issued by the Bird Banding Lab of the US Geological Survey (USGS). On the kite's other leg, she put a green band with "V2" stamped into it. This one is for the UF Snail Kite Monitoring Program. Resighting banded birds helps the monitoring program estimate population size.

Kent gently withdrew the kite's tail feathers from the tube, inch by inch. The feathers at the mouth of the tube exhibited the iconic white patch, followed by a thick band of brown, and a curve of cream along the end of each feather. Kent spread the feathers, sorted through them, and measured them.

Making constant adjustments to the feathers and the wings to ease the experience for the bird, Kent gradually pulled the kite the rest of the way out.

As the kite's head emerged from the tube, Bouchenot held her up. We could see her alarm in her open bill, bright orange at its base and tipped with black. I was startled to see her bright-red eyes up close.

To calm her, Kent fitted a little hood on her, a practice borrowed from falconry. It was a puffy pale-green fabric helmet with a fanciful green tassel on top. Below the hood I could see her bill, and the soft feathers of her neck.

Bouchenot invited me to touch an area a bit below her neck. There was a lump. Bouchenot grinned and widened her eyes, whispering, "That's her lunch." She told me that when snail kites eat a snail, at first it's in this part of them, called the "crop," like an upper stomach.

Next, she invited me to feel the breastbone, which turns out to be a wire of a bone with sharp little knobs on it. Nothing like the robust chicken breasts I've handled. As Kent prepared her materials, she and Bouchenot whispered speculation that this kite may have had a brood this nesting season, so she may be a little skinnier from prioritizing snails for her babies. Kent said she was not the skinniest snail kite she had handled. She weighed 500 grams (1.1 pounds).

Bouchenot whispered that it cracks her up how docile kites are. In her experience, people handling birds of prey need to be trained to protect themselves from the thrashing of dangerous beak and talons. This kite was perfectly still in her hand. It seems that when caught and capped with a hood, snail kites go into "freeze" mode rather than fight.

At Kent's direction, Bouchenot lay the kite down on the back of the truck. Kent deftly took the necessary measurements and samples, not rushing, but not sparing a second. I was mesmerized. The two scientists stretched out one wing at a time, counting and measuring distinct types of feathers. They both whispered, "Aww," when they saw one of her wing feathers was a very short newly growing feather. The clear tube that is at the base of each feather was darker and much longer for this one. Bouchenot whispered to me that lots of blood and nutrients were being delivered to the end of that tube, and that was where the shiny blackish brown feather was growing from. I had never known that was how feathers grew. Kent said it's called a blood feather, or a pin feather.

Gaining so much inside knowledge so quickly, while watching people whispering through strange rituals, I felt like I was being inducted into a secret society.

The wings were revelatory as well. The only nonfeathered parts of each wing were the articulated bones that ran along the upper edge. I did not realize that the flesh and bone of a raptor's wing was much like a curtain rod, where the curtain is simply dangling feathers. Indeed, most of the breadth of the wing was a single layer of long and stout feathers, up against each other to give the illusion of a continuous surface.

When it was time to put the GPS transmitter on the kite, Kent asked me to hold her. Bouchenot transferred her into my hands, demonstrating how to place my fingers to support her legs and body.

I held her body with my right hand, while my left hand held her hood on. Kent eased the straps around her wings and gently adjusted it to secure the lightweight transmitter like a miniature backpack.

Kent ushered me to the shady side of the truck and whispered to me, "Lower your hands." She had to repeat it because my brain was not functioning. I existed solely to hold this snail kite in my hands. Awash with sensory input, I had no thoughts. The most prominent feeling was that her body was warm and light. As big as kites are in flight, I had not expected one to feel so slight, like a fistful of feathers. Her chest feathers were silky and soft. Up close I was struck by how perfectly each feather was shaped and colored, each one a work of art, mostly brown and black, with little white wispy feathers close to her skin.

As Kent worked with great concentration on the harness, I put all my attention into my fingers. I slowed my breathing and got very still in an effort calm the rapid beating of my heart. I tried in vain to feel the rise and fall of her breath with my hand. Later I learned why it eluded me. Birds have circular breathing, lacking the tides of inhalation and exhalation we mammals have. I shifted a finger up a little higher toward her chest and thought I could feel a quicker movement. Could that be her heartbeat? I couldn't be sure.

I inhaled deeply through my nose. I realized that I had expected that she would smell like a chicken coop, but she barely smelled like anything. A faint earthy smell.

Kent secured the harness where the straps crossed in the front. She gently adjusted it and finally decided that it would stay in place without being tight.

And just like that, it was time for goodbye.

Kent gently took the kite from me and carried her several steps away from the truck. Bouchenot and I followed. Kent held her away from her body, to make room for the flapping of wings. She adjusted her hands so that she was not holding the wings. Bouchenot came from behind and put her hand on the top of the hood.

A pause.

Then Bouchenot lifted off her hood.

And there the kite's face was again with her brilliant-red eyes, alive and wild. She looked down at first, raising her wings up as a human would raise their elbows above their ears. Then she raised the tips of her wings all the way up. Looking ahead, she flapped her wings down and back up, several times, each time extending them more. Kent reached her arms farther out to stay out of the way of the rapidly lengthening wingspan. The snail kite was growing before our eyes, larger than I thought possible for the slight body that had nestled so easily in my hand.

Suddenly, with a sequence of audible flaps, the kite surged forward and upward out of Kent's uplifted hands.

We watched her fly off above the marsh where we had spent so much time hoping and watching and waiting to have this close encounter with her.

Kent said to her, "Thank you my love, that was beautiful!" As she became a dot, Kent added with a big smile, "Bye, baby, thank you!"

Kent checked her watch. From the time I had texted her "hovered," it had been fifty minutes.

I congratulated her, and she congratulated me back. She gave me a big hug while jumping up and down a bit. The three of us laughed and broke out into a little happy dance.

We packed up, chatting about what had just happened.

"I'm still shaking!" Kent said. Through the celebratory atmosphere, I could see how stressful capturing and handling a wild bird was for Kent. Her dedication to helping the species showed in the long, focused hours she put in over multiple trips, her ability to dive in with surgical precision to minimize time and discomfort for the captured bird, and the utmost respect and care she gave the kite at every moment. Kent had been almost as eager as the bird for it to be free.

FIGURE 28.1. Gina Kent (*right*) prepares to let go of a female snail kite after taking samples, attaching a GPS harness, and banding its ankles, as part of her work with Avian Research and Conservation Institute (ARCI). Jenny Bouchenot (*left*) is ready with a plastic lid to catch any bird droppings for scientific analysis. Photo by Hilary Flower.

In the truck as Kent was starting to drive us back to the parking lot where the day had started, she said we needed to name her. The GPS-tagged kites get a two-part name. The first is where it was caught; this spot was called ROMA. The other two kites that Kent had caught here were ROMA Uno and ROMA Dos. She said we could call her ROMA Tres, but maybe we could come up with something better.

I said, "What about Red? For her bright-red eyes?"

Bouchenot said, "Like a Roma tomato!"

And it stuck.

I knew it would be a long time before I fully absorbed what I had experienced. After parting with Kent and Bouchenot at the parking lot, I started my long drive for home through the sunset with almost no thoughts in my head. It was a little like being in shock. I just kept feeling her warm feathers in my hand.

The next day I got a text from Kent. "Red went back to the communal roost and came back to her foraging area this morning."

Somehow, I had not stopped to think that Red would stay in touch with us. I was tickled to know that she had gone back to the communal roost, and was foraging again where we had seen her, probably on the little cypress tree or the lone cypress tree. I wondered where she would be this time next year. I looked forward to following her as dots on a map, at least for the two years or so that GPS trackers last.

For days, I could still feel her warm, feathery form in my right hand. I thought about how I had set off on a quest for hope over a year ago, and I had chosen the snail kite as my guide. And I had been granted my wish to hold one.

And it had been a revelation.

For the minutes of holding her, all my complicated ideas about the snail kites and handwringing about their future had fallen away.

When you're holding a snail kite, I realized, hope is the only option. It's not something to find, or something that can be lost. It's warm, soft, and alive in your hands. Like Kent said, you have to hope.

I smiled as I recalled that Jean Takekawa's response when I had asked if she had hope, was the same as Kent's: "I have to." Emily Dickinson knew: "Hope is the thing with feathers."

Now I understood.

To hold in your hands this living, breathing creature, so improbable, so fragile, so capable.

I had held hope in my hands.

# ACKNOWLEDGMENTS

I am grateful to my children: to Sal, for pointing out my first snail kite and countless other birds, and for being a source of insight and support throughout; to Dervin, for illustration advice, decision-making-guidance, and cat sitting while I was in the field; to Ramsay, for being my on-call IT support and for teaching me the word "milvine" (kite-like). I am grateful to my late father, Daniel, who loved to collect the large shells of the invasive snail, for showing me the writer's life, and to my mother, Kate, for always being on my team.

My editor Janie Chan shared my vision for this project from the start and helped shape it into the best version it could be. Emily Hunsaker helped revise chapter 1, Sian Hunter helped the book in the production and publication phases, and Marthe Walters helped with editing, design and production. I am also grateful to three anonymous peer reviewers, each of whom made the book substantially better. One invested an extraordinary amount of time and taught me a lot about writing.

I am grateful to all of my friends who gave moral support and even read some or all of the manuscript. Adam Bailey and Lee Irby helped me with the earliest stages of the project, believing in it and reading the first chapters. I am deeply grateful to Kait Kennedy, Dr. Maureen Corbett, Karen Corbett, Linda Theilman, Annette Reiter, and Julie Menke for making it through the whole manuscript, and making it better. Laura Josler and Dr. Andrew Chittick both gave feedback on early chapters. I am grateful to my friends who encouraged me and offered ideas, including Bindu Usher, Dr. Eileen Otis, China Martin, Dr. Teresa Gaskill, and Shelly Tiernan. Shane Tiernan, who needs to get his own books out into the world, made me laugh. Liesl Piotti restored me to sanity. Max Francois ensured that I was always "here to learn and receive."

Eckerd College has been a big part of this book on many levels. I am grateful to former students Kait Kennedy and Zoe Sabadish for accompanying me on snail kite adventures and bringing their perspectives into the book. My colleague and friend Dr. Beth Forys let me pick her brains about the world of ornithology and gave me key feedback on several chapters. Dr. Jo Huxster came along for all of the ups and downs; I am excited that she now has embarked on her own book adventure. Thank you to my writing partner Dr. Lisa Miller, who also has a book in the works! Dr. Suzan Harrison and Dr. Amanda Hagood and their writing students were generous readers of chapter 2.

Eckerd students Jordan Denis and Yeshvant Gill helped with Chip. Eckerd College hosts an annual Writers in Paradise conference, which I participated in. Shout-out to Madeleine Blais for helping me polish the first chapter, and for the nonfiction workshop she led; I am grateful to workshop participants Rosanne Annoni, Darla Chesnet, Marie Corbett, Charmaine Denison-George, Evelyn Krieger, Ellen Leesfield, Marah Reinoso Vega, Chris Teare, Mary Jane White, and Ted Williams. Thank you to Eckerd College for offering financial support for my field travel, for faculty writing retreats, and for counting fun projects like this as professional productivity.

When I first started reaching out to strangers with kite expertise, it never occurred to me that some of them would become such untiring supporters of me and the project. I cannot bear to think of this book without the contributions of Dr. Paul Gray, Dr. Phil Darby, Dr. Ken Meyer, Dr. Steve Beissinger, and Jean Takekawa. Getting to know you all is what I most treasure about this project. Dr. Paul Gray was a huge part of this project from the beginning to the very last minute, humoring countless questions ("You again!"), providing detailed notes on the whole manuscript, and going back and forth with me to get the hardest sections right. Thank you, Paul, for taking me out to Indian Prairie and sharing your snail kite world with me. Dr. Phil Darby shared his expertise in multiple interviews, helped me get his sections right, brought me to his discovery on Lake Kissimmee, and gave me moral support. More than this, Phil Darby made me see the native apple snail and the Everglades in a whole new light. Dr. Ken Meyer was a font of passion and insight, and an inspiration to never stop fighting for imperiled animals. I am grateful to him for sharing his hard-won understanding of what snail kites need from us. Dr. Steve Beissinger did some heavy lifting for this project, always cheerfully answering my endless questions in emails and interviews. A legend in the world of snail kite research, Steve gave me the best snail kite education anyone could hope for. Jean Takekawa quickly became the heart of this book. She shared the wonder of her experiences with me and showed me what a life of conservation can achieve. She has left her mark on the world of snail kites and many more wild animals and wild places. I am also grateful to her brother John Takekawa and Steve Beissinger for helping me reach her. Jean wishes to credit Joyce Kleen, Ken Snyder, and Carol Snyder with helping her accomplish the 372-kite roost count in 1985.

I am deeply honored that Miccosukee Elder Michael Frank shared himself with me and this project, allowing me into his Everglades home and generously imparting his experiences and wisdom. Thank you to Amy Castaneda, Dr. Marcel Bozas, Ed Ornstein, and Dr. Jason Daniels for interviewing with me and helping me get chapter 25 right.

I had never heard of Grassy Waters Preserve before this project; it was one of the big surprises for me, and an inspiration for what we can do for snail kites and wildlife. I am grateful to Lauren Butcher, Vera de Chalambert, and Chloe Carter for sharing the wonders of Grassy Waters with me, and to the City of West Palm Beach for taking such good care of it. Another huge surprise was the beautiful 372-kite roost amid a solid waste facility. I am grateful to David Broten and Mandy Krupa for showing me this ongoing miracle. They help the Solid Waste Authority of Palm Beach County take good care of this crucial roost and wading bird rookery. I must extend a special thank-you to Lauren Butcher and David Broten for thoughtful feedback on their chapters.

I am grateful to Dr. Caroline Poli, Gina Kent, Eric Crawford, Tyler Beck, Matt Pearce, Andrea Leavitt Anderson, Dr. Phil Darby, Dr. Paul Gray, Nathan Barrus, Lauren Butcher, David Broten, Dr. Kai Rains, Dr. Mark Rains, and Jennifer Brinkworth for sharing unforgettable field opportunities with me while imparting their expertise.

Dr. Wiley Kitchens, Dr. Brian Reichert, and Dr. Zach Welch invested significant time to share stories and knowledge, and to help me get their chapters right. Jaclyn Lopez was very patient with me as I struggled to understand the Endangered Species Act (ESA). Steve Schubert gave me a much-needed view into the practices and culture at the US Fish and Wildlife Service (USFWS) related to the ESA, and spent a lot of time working with me to get it right. Jane Tutton also gave me an invaluable perspective on the USFWS. Thank you to Dr. Paul Sharp, Dr. Tim Collins, and Dr. Phil Darby for sharing the results of their parasite study with me. I am grateful to Holly Andreotta, Dr. Paul Sykes, Dr. Noel F. R. Snyder, S. Ansley Samson, Dr. Peter Grant, Dr. Rosemary Grant, Dr. Ary Hoffman, Dr. Joel Trexler, Dr. Paul Julian, Hanna Innocent, Tommy Shannon, Dr. Rob Fletcher, Brian Jeffery, Dr. Kathryn Smith, Travis Thompson, Dave Mellow, Dr. Jacoby Carter, Melissa Tolbert, Dr. Lance Gunderson, Dr. Julien Martin, Dr. Darren Rumbold, and Manley Fuller for interviewing with me for this book. Getting to know the many people whose work involves snail kites or the wetlands they depend upon was the true adventure of this book.

Dr. Mark Rains took me on my first visit to the Everglades and made me an Everglades ecohydrologist. He taught me how the world of wetlands works, from the aquifer up to the lofty heights of government. Dr. Mark Rains and Dr. Kai Rains have supported me and this project in countless ways.

I am indebted to Tim Barker (www.timbarkerphotography.com) for the cover photo and other snail kite photos in this book. Thank you also to John Duncan (www.johnduncanphotography.com) for sharing snail kite photos for promoting the book.

Dr. Julie Armstrong, Dr. Tom Hallock, Lee Irby, Craig Pittman, K. C. Wolfe, and Eric Deggans generously shared with me their guidance and insights on book writing and publishing. Tanya Coovadia, who did not live long enough to write all of the books she had inside of her, was always certain that I had an Everglades book inside of me.

# GLOSSARY OF ACRONYMS AND INFORMAL NOMENCLATURE

3A Water Conservation Area 3A
ARCI Avian Research and Conservation Institute
EAA Everglades Agricultural Area
ENP, or "the Park" Everglades National Park
ERM Environmental Resource Management of Palm Beach County
ESA Endangered Species Act
FDEP, or DEP Florida Department of Environmental Protection
FIU Florida International University, Miami
FWC Florida Fish and Wildlife Conservation Commission
NRCS Natural Resources Conservation Service
SFWMD, or "the District" South Florida Water Management District
SSA Species Status Assessment
STA Stormwater Treatment Area
SWA Solid Waste Authority
UCF University of Central Florida, Orlando
UF University of Florida, Gainesville
USACE, or "the Corps" United States Army Corps of Engineers
USDA United States Department of Agriculture
USFWS, or "the Service" United States Fish and Wildlife Service
USGS United States Geological Survey
UWF University of West Florida, Pensacola
WCA Water Conservation Area
WCA-1, or "the Refuge" Arthur R. Marshall Loxahatchee National Wildlife Refuge

# NOTE ON SOURCES AND SELECTED BIBLIOGRAPHY

I first learned about snail kites reading the excellent chapter on them in Ted Levin's *Liquid Land* (University of Georgia Press, 2004).

To research this book, I interviewed more than sixty sources in person and via video call, phone, and email. Many sent me key reports, letters, clippings, photos, presentations, scientific articles, and contact info for other experts.

Five individuals served as ongoing sources of information and interpretation throughout this project. Through repeated conversations and generous sharing of expertise, stories, insights, and archival materials, they informed many chapters and shaped the book's factual foundation, even in places where their names do not appear. Each of them read and gave notes on sections pertaining to their work.

- Dr. Paul Gray has advocated for wildlife in the Everglades for decades. He was a graduate student in the late 1980s, worked for the Florida Fish and Wildlife Conservation Commission (FWC) in the early 1990s, and has been with Audubon Florida since 1995. He is currently the Everglades science coordinator for Audubon Florida. He took me out to Indian Prairie. He gave detailed notes on the entire manuscript, and subsequent drafts of chapters 10 and 17 through 21.
- Dr. Steve Beissinger wrote breakthrough studies on snail kites with fieldwork from 1979 through the 1980s. He is still involved in snail kite research today. He is professor emeritus at the University of California at Berkeley.
- Dr. Phil Darby is the expert on all things apple snail in Florida. He has researched and published on snail kites and apple snails for thirty years, initially at UF. He took me out to see the burgeoning popula-

tion of native apple snails in Lake Kissimmee. He is professor at the University of West Florida, where he teaches ecology and leads a research lab.

- Jean Takekawa did key conservation work and research with snail kites in the 1980s as a wildlife biologist for the US Fish and Wildlife Service (USFWS) at Loxahatchee National Wildlife Refuge. She discovered and helped save many emergency wetlands for snail kites. After leaving Florida, she went on to manage the Nisqually National Wildlife Refuge. She gave notes on the entire manuscript.
- Dr. Ken Meyer cofounded the Avian Research and Conservation Institute (ARCI) in 1997. As director of ARCI he has gathered scientific data to inform conservation for snail kites and many other imperiled birds in the United States and Latin America.

This project benefited from the perspectives and experiences of a host of other experts, each of whom made important contributions to the book.

Gina Kent is the senior conservation scientist at the Avian Research and Conservation Institute (ARCI), where she has been researching and doing hands-on conservation work with snail kites and imperiled birds with Dr. Ken Meyer since 2001. Gina Kent gave me invaluable field experiences with snail kites, and gave me notes on chapters 4, 27, and 28. Melissa Tolbert is the environmental program supervisor for the Loxahatchee Slough Natural Area for Palm Beach County's Department of Environmental Resources Management. She helped turn abandoned agricultural lands into a thriving ecosystem enjoyed by snail kites and countless wading birds. She is an important partner facilitating the work of ARCI. Jennifer Bouchenot is a PhD student at the University of Central Florida in Orlando, studying contaminants and raptors. She is collaborating with ARCI to determine the methylmercury exposure of the snail kites that Kent traps. I also met with staff from the Palm Beach Zoo and Conservation Society who support ARCI's work.

Conversations with some of the early snail kite researchers added essential historical context. Dr. Paul Sykes did key snail kite studies in the 1960s through the 1980s as a biologist for the USFWS, at the Patuxent Wildlife Research Center, at Delray Beach Station in Florida. Among other things, he told me about his airboat breaking down out in the middle of 3A, and his having to spend the night out there before wading several miles out to the canal in the morning. Dr. Noel F. R. Snyder conducted fieldwork with snail kites for the USFWS and coauthored key papers on them in the 1970s and 1980s. He is a renowned raptor specialist; with Helen Snyder he cowrote *Raptors of North America: Natural History and Conservation.* Dr. Gary Falxa was a field tech studying snail kites for the 1979 season with Beissinger under Snyder. He worked as a wildlife biologist for the USFWS for twenty years. Dr. Snyder generously offered photos of his fieldwork with Steve Beissinger and Gary Falxa (figure 6.1, figure 6.2, and figure 6.3).

The snail kite monitoring program at the University of Florida (UF) is a huge source of knowledge and conservation guidance. I spoke with several researchers, past and present. Dr. Wiley Kitchens directed the program for many years; he is now professor emeritus at UF. He put me in touch with many of his former field crew and graduate students. Dr. Brian Reichert witnessed the snail kites' transition to the invasive snail on Lake Toho, initially as a field tech before earning his master's and PhD. He is currently the branch chief of Ecoinformatics & Wildlife Technology at the US Geological Survey Fort Collins Science Center, in Colorado. Dr. Julien Martin wrote his PhD dissertation on snail kites; he is currently at the US Geological Survey, Eastern Ecological Science Center in Laurel, Maryland. Dr. Zach Welch wrote his master's thesis on the 2004 restoration project at Lake Toho that launched the spread of the snail *P. maculata.* He started as a field tech and completed a PhD and postdoctoral research fellowship on snail kites. He was the snail kite coordinator for the FWC from 2009 to 2014. Dr. Welch is a section administrator for South Florida Water Management District. Dr. Kitchens, Dr. Reichert, and Dr. Welch gave me notes on chapters relating to their work.

Dr. Robert Fletcher was Dr. Kitchens's successor as director of the snail kite monitoring program at UF. Dr. Caroline Poli was his PhD student and postdoctoral researcher on snail kites; she showed me the snail kites at Paynes Prairie and discussed her research. Brian Jeffery was the project manager for the program and Dr. Fletcher's PhD student.

I spoke with several people with expertise in the implementation of the Endangered Species Act in Florida. S. Ansley Samson is the general counsel at Everglades Law Center, Inc. Jaclyn Lopez was the Florida director and senior attorney for the Center for Biological Diversity for over a decade. She is currently assistant professor of law and director of the Jacobs Public Interest Law Clinic for Democracy and the Environment at Stetson Law School in St Petersburg, Florida. Tyler Beck has been the snail kite conservation coordinator at the Florida Fish and Wildlife Conservation Commission for more than a decade. He gave me a tour of Moonshine Bay to observe snail kite nesting in an unusual habitat treatment. Steve Schubert was a fish and wildlife biologist at the USFWS Vero Beach office from 2000 to 2020, becoming involved with snail kites in 2006, among other species. He wrote the 2018 Biological Opinion on snail kites at Lake Okeechobee and consulted on Everglades Restoration. He is now the biological programs manager at Earthology Consulting Services, LLC. Schubert worked closely with me to develop and revise chapter 21. Jane Tutton was the endangered species coordinator at the USFWS Vero Beach office from 1991 to 2014. Dr. Mark Rains is the chief science officer, Florida Department of Environmental Protection, and professor of ecohydrology in the School of Geosciences at the University of South Florida.

I interviewed Eric Crawford about the impacts of *P. maculata* on vegetation, and he showed me the location of the 2013 invasive snail boom. He gave me notes on chapters 12 and 13. He is the lead scientist for Stormwater Treatment Area (STA) Vegetation Management at the South Florida Water Management District (SFWMD), where he has worked for seventeen years. I spoke with Holly Andreotta about the impacts of snail kites on STA management; she is the lead environmental analyst at the SFWMD.

Chapter 25 draws on interviews and field observations with several representatives of the Miccosukee Tribe of Indians of Florida, who shared information and perspectives on their concerns about the degradation of Water Conservation Area 3A. Michael Frank is an elder of the Tribe. Ed Ornstein is the Tribe's deputy general counsel. Dr. Marcel Bozas is the director of the Fish and Wildlife Department. He wrote his PhD dissertation at Florida International University on mammals' use of tree islands in 3A. Amy Castaneda is the water resources director. Dr. Jason Daniel is the Tribal historic preservation officer. Mr. Frank, Mr. Ornstein, Dr. Bozas, and Ms. Castaneda all read and gave notes on chapter 25; Dr. Daniel helped improve key sections. Representatives from the Tribe shared with me three sources that appear to support the idea that the Miccosukee Tribe integrated with remnants of other Indigenous groups in South Florida: (1) Patsy West, "Abiaka, or Sam Jones, in Context: The Mikasuki Ethnogenesis through the Third Seminole War," *Florida Historical Quarterly* 94, no. 3 (2016): 366–410; (2) John Worth, "Creolization in Southwest Florida: Cuban Fishermen and 'Spanish Indians,' ca. 1766–1841," *Historical Archaeology* 46, no. 1 (2012): 142–60; and (3) David Rahehe-tih Webb, *The Spanish Seminole* (Florida Historical Society Press, 2023), 1.

I visited and spoke with several staff members of the Grassy Waters Preserve, City of West Palm Beach. Lauren Butcher is the environmental education coordinator; she gave me a kayak tour of the preserve and provided feedback on chapter 9. Chloe Carter and Vera de Chalambert are both educators.

I also had the opportunity to observe the snail kite roost at the Solid Waste Authority (SWA) of Palm Beach County and speak with many people who have done important work on it. Dr. Darren Rumbold monitored snail kites there from 1987 to 1991. He is professor of marine science at Florida Gulf Coast University. David Broten leads the snail kite counts for the SWA, as the environmental programs manager. Mandy Krupa works there and has assisted with the counts for more than fourteen years.

I spoke with Dr. Evelyn Gaiser and two of her graduate students at Florida International University (FIU). Hanna Innocent earned her master's studying periphyton in Lake Okeechobee. Tommy Shannon is earning his PhD. Nate Barrus wrote his master's thesis on *P. maculata* and *P. paludosa* under Dr. Nate Dorn at FIU. He is now continuing this work as a PhD candidate.

Dr. Beth Forys is professor of environmental science and biology and the Richard R. Hallin Endowed Chair of Natural Sciences at Eckerd College in St. Petersburg, Florida. She was an ongoing source of ornithology knowledge and insight, and gave notes on several chapters.

Dr. Kathryn Smith wrote the Species Status Assessment for the snail kite, as the project manager at Texas A&M Natural Resources Institute on the Strategic Conservation Policy and Planning team.

Dave Mellow discovered and studied the early explosion of the snail *P. maculata* at Lake Toho in 2004, as part of his work as the field tech for Dr. Phil Darby of the University of West Florida.

Information about the 2023 snail kite nesting event at Pearce Ranch came through conversations with Matt Pearce, an eighth-generation Florida cattle rancher and owner of Pearce Ranch, and Travis Thompson, the executive director of All Florida. Mr. Pearce showed me the location of the snail kite nests at the ranch. Mr. Pearce and Mr. Thompson worked with the UF snail kite monitoring program to care for and document the nesting event. Both spoke to me about the relationship between cattle ranching and snail kite conservation.

This book features insights from correspondence with leading researchers in the field of evolutionary biology. Dr. Rosemary Grant and Dr. Peter Grant, evolutionary biologists at Princeton University (now emeritus), are renowned for their groundbreaking, long-term studies of Darwin's finches on the Galápagos Islands, which started in 1973. Their research has provided crucial evidence for climate-driven rapid evolution. In interviews, they discussed how species can rapidly adapt to environmental changes. Also contributing to this line of inquiry was Dr. Ary Hoffman, chair of Ecological Genetics at the University of Melbourne, Australia.

Dr. Lance Gunderson, professor and chair of the Department of Environmental Sciences at Emory University in Atlanta, Georgia, spoke with me about the dynamic nature of ecosystems and the need for adaptive management. He worked for a decade for the US National Park Service in Big Cypress Preserve and the Everglades National Park. He and C. S. Holling coined the term "panarchy" in their 1995 book *Barriers and Bridges to the Renewal of Ecosystems and Institutions.*

Dr. Tim Collins is professor emeritus of biological sciences at FIU, and Dr. Paul R. Sharp is assistant director of biological sciences at FIU. They generously shared the results of the parasite study they conducted with Dr. Phil Darby on *P. paludosa* and *P. maculata* in Florida, which is as yet unpublished.

Dr. Jacoby Carter spoke with me about range expansion for limpkins, snails, and snail kites from his experience, but not as a representative of any agency. Dr. Carter gave notes on his section of chapter 26. He was a research ecologist at the US Geological Survey, Wetland and Aquatic Research Center in Lafayette, Louisiana, retired in 2003. Manley Fuller spent thirty-two years as president and CEO for Florida Wildlife Federation (FWF). He is now vice president of conservation policy at the North Carolina Wildlife Federation. He tried in vain to find information on the 2005 lawsuit brought on behalf of the snail kites by the National Wildlife Federation and the Florida Wildlife Federation. Dr. Paul Julian is the biogeochemist for the Everglades Foundation; he spoke to me about Everglades restoration. Dr. Joel Trexler has studied apple snails and other wildlife in the Everglades. He is the director of the Coastal and Marine Laboratory in St. Teresa, Florida, where he is professor of biological science. Jennifer Brinkworth brought me out to observe snail kites at Lake Toho as part of her consulting work as staff environmental scientist and avian biologist at SWCA Environmental Consultants, Jacksonville, Florida. Andrea Leavitt Anderson was a ranger at Boyd Hill Nature Preserve in St Petersburg, Florida, for many years. She was there for a full boom-and-bust cycle of *P. maculata.*

I relied on numerous peer-reviewed scientific journal articles, especially:

Beissinger, S. R. "Demography, Environmental Uncertainty, and the Evolution of Mate Desertion in the Snail Kite." *Ecology* 67, no. 6 (1986): 1445–59.

Beissinger, S. R., and J. E. Takekawa. "Habitat Use by and Dispersal of Snail Kites in Florida During Drought Conditions." *Florida Field Naturalist* 11 (1983): 1.

Cattau, C. E., R. J. Fletcher Jr., R. T. Kimball, C. W. Miller, and W. M. Kitchens. "Rapid Morphological Change of a Top Predator with the Invasion of a Novel Prey." *Nature Ecology & Evolution* 2, no. 1 (2018): 108–15.

Cattau, C. E., R. J. Fletcher Jr., B. E. Reichert, and W. M. Kitchens. "Counteracting Effects of a Non-Native Prey on the Demography of a Native Predator Culminate in Positive Population Growth." *Ecological Applications* 26, no. 7 (2016): 1952–68.

Cattau, C. E., J. Martin, and W. M. Kitchens. "Effects of an Exotic Prey Species on a Native Specialist: Example of the Snail Kite." *Biological Conservation* 143, no. 2 (2010): 513–20.

Darby, P. C., I. Fujisaki, and D. J. Mellow. "The Effects of Prey Density on Capture Times and Foraging Success of Course-Hunting Adult Snail

Kites." *The Condor* 114, no. 4 (2012): 755–63.

Darby, P. C., D. J. Mellow, and M. L. Watford. "Food-Handling Difficulties for Snail Kites Capturing Non-Native Apple Snails." *Florida Field Naturalist* 35, no. 3 (2007): 2.

Gutierre, S. M., P. C. Darby, P. L. Valentine-Darby, D. J. Mellow, M. Therrien, and M. L. Watford. "Contrasting Patterns of *Pomacea maculata* Establishment and Dispersal in an Everglades Wetland Unit and a Central Florida Lake." *Diversity* 11, no. 10 (2019): 183.

O'Neil, Chase M., Yuxi Guo, Steffan Pierre, Elizabeth H. Boughton, and Jiangxiao Qiu. "Invasive Snails Alter Multiple Ecosystem Functions in Subtropical Wetlands." *Science of the Total Environment* 864 (2023): 160939.

Poli, C., E. P. Robertson, J. Martin, A. N. Powell, and R. J. Fletcher Jr. 2022. "An Invasive Prey Provides Long-Lasting Silver Spoon Effects for an Endangered Predator." *Proceedings of the Royal Society* B 289 no. 1977 (2022): 20220820.

Rumbold, Darren G., and Mary Beth Mihalik. "Snail Kite Use of a Drought-Related Habitat and Communal Roost in West Palm Beach, Florida 1987–1991." *Florida Field Naturalist* 22, no. 2 (1994): 1.

Snyder, N. F., S. R. Beissinger, and R. E. Chandler. "Reproduction and Demography of the Florida Everglades (Snail) Kite." *The Condor* 91, no. 2 (1989): 300–316.

Sykes P. W., Jr. "The Range of the Snail Kite and Its History." *Bulletin of the Florida State Museum: Biological Sciences* 29 (1984): 211.

Takekawa, J. E., and S. R. Beissinger. "Cyclic Drought, Dispersal, and the Conservation of the Snail Kite in Florida: Lessons in Critical Habitat." *Conservation Biology* 3, no. 3 (1989): 302–11.

An extraordinary reference work that I learned a great deal from is a 2013 peer-reviewed report written principally by Patricia Valentine-Darby. Often referred to as the Pomacea Project, this report is a master class in literature review covering everything you could want to know about the snail kite world up to 2013, including comprehensive information on the native and invasive apple snails. The report is: Pomacea Project, *Literature Review of Florida Apple Snails and Snail Kites, and Recommendations for their Adaptive*

*Management. Final Report.* 2013. Submitted to the National Park Service, Everglades National Park, by the Pomacea Project, Inc., Pensacola, FL.

For nesting data and kite population estimates from 2002 to 2024, I drew on reports from the snail kite monitoring program at UF (beginning with the most recent):

Fletcher, Robert, Caroline Poli, Brian Jeffery, Meghan Beatty, Lara Elmquist, and Miguel Acevedo. *Snail Kite Demography: 2024 Annual Report on the 2023 Breeding Season.* Report prepared for U.S. Army Corps of Engineers. 2024.

Fletcher, Robert, Caroline Poli, Brian Jeffery, Meghan Beatty, and Lara Elmquist. *Snail Kite Demography: 2023 Annual Report on the 2022 Breeding Season.* Report prepared for US Army Corps of Engineers. 2023.

Fletcher, Robert, Caroline Poli, Brian Jeffery, Meghan Beatty, and Alfredo Gónzalez. *Snail Kite Demography: 2022 Annual Report on the 2021 Breeding Season.* Report prepared for US Army Corps of Engineers. 2022.

Fletcher, Robert, Caroline Poli, Brian Jeffery, Meghan Beatty, and Alfredo Gonzalez. *Snail Kite Demography: 2021 Annual Report on the 2020 Breeding Season.* Report prepared for US Army Corps of Engineers. 2021.

Fletcher, Robert, Caroline Poli, Brian Jeffery, Brian Reichert, Friederike Potash, Ellen Robertson, and Alfredo Gonzalez. *Snail Kite Demography: 5-Year Final Report and Update on the 2019 Breeding Season.* Report prepared for US Army Corps of Engineers. 2020.

Fletcher, Robert, Ellen Robertson, Sarah Dudek, Caroline Poli, and Brian Jeffery. *Snail Kite Demography: 2019 Annual Report on the 2018 Breeding Season.* Report prepared for US Army Corps of Engineers. 2019.

Fletcher, Robert, Ellen Robertson, Brian Jeffery, Caroline Poli, and Sarah Dudek. *Snail Kite Demography: 2018 Annual Report on the 2017 Breeding Season.* Report prepared for US Army Corps of Engineers. 2018.

Fletcher, Robert, Caroline Poli, Ellen Robertson, Brian Jeffery, Sarah Dudek, and Brian Reichert. 2017. *Snail Kite Demography: Annual Progress Report for the 2016 Breeding Season.* Report prepared for US Army Corps of Engineers. 2017.

Fletcher, Robert, Ellen Robertson, Caroline Poli, Brian Jeffery, Brian Reichert, and Christopher Cattau. *Snail Kite Demography: 2015 Annual Report.* Report prepared for U.S. Army Corps of Engineers. 2016.

Fletcher, Robert, Ellen Robertson, Brian Reichert, Christopher Cattau, Rebecca Wilcox, Christa Zweig, Brian Jeffery, Jean Olbert, Kyle Pias, and Wiley Kitchens. *Snail Kite Demography: 5-Year Report, Final Report 2014.* Report prepared for U.S. Army Corps of Engineers. 2015.

Fletcher, Robert, Christopher Cattau, Rebecca Wilcox, Christa Zweig, Brian Jeffery, Ellen Robertson, Brian Reichert, and Wiley Kitchens. *Snail Kite Demography: Annual Progress Report 2013.* Report prepared for U.S. Army Corps of Engineers. 2014.

Cattau, Christopher, Brian Reichert, Wiley Kitchens, Robert Fletcher, Jean Olbert, Kyle Pias, Ellen Robertson, Rebecca Wilcox, and Christa Zweig. *Snail Kite Demography: Annual Report 2012.* Report prepared for U.S. Army Corps of Engineers. 2012.

Reichert, Brian, Christopher Cattau, Wiley Kitchens, Robert Fletcher, Jean Olbert, Kyle Pias, and Christa Zweig. *Snail Kite Demography: Annual Report 2011.* Report prepared for U.S. Army Corps of Engineers. 2011.

Reichert, Brian, Christopher Cattau, Wiley Kitchens, Robert Fletcher, Jean Olbert, Kyle Pias, Christa Zweig, and Jeremy Wood. *Snail Kite Demography: Annual Report 2010.* Report prepared for U.S. Army Corps of Engineers. 2011.

Cattau, Christopher, Wiley Kitchens, Brian Reichert, Jean Olbert, Kyle Pias, Julien Martin, and Christa Zweig. *Snail Kite Demography: Annual Report 2009.* Report prepared for U.S. Army Corps of Engineers. 2009.

Cattau, Christopher, Wiley Kitchens, Andrea Bowling, Brian Reichert, and Julien Martin. *Snail Kite Demography: Annual Report 2007.* Report prepared for U.S. Fish and Wildlife Service. 2008.

Martin, Julien, Wiley Kitchens, Christopher Cattau, Andrea Bowling, Sara Stocco, Eric Powers, Christa Zweig, Althea Hotaling, Zach Welch, Hardin Waddle, and Alejandro Paredes. *Snail Kite Demography: Annual Progress Report 2006.* Report prepared for U.S. Fish and Wildlife Service. 2007.

Martin, Julien, Wiley Kitchens, Christopher Cattau, Andrea Bowling, Melinda Conners, and Daniel Huser. *Demography of the Snail Kite in*

*Blue Cypress Marsh Complex: Final Report 2005*. Report prepared for St. Johns River Water Management District. 2005.

Martin, Julien, Wiley Kitchens, Christopher Cattau, Christina Rich, and Derek Piotrowicz. *Snail Kite Demography: Annual Report 2004*. Report prepared for US Fish and Wildlife Service. 2004.

Martin, Julien, Wiley Kitchens, and Michaela Speirs. *Snail Kite Demography: Annual Report 2003*. Report prepared for US Fish and Wildlife Service. 2003.

Martin, Julien, Zachariah Welch, Samantha Musgrave, Derek Piotrowicz, and Wiley Kitchens. *Snail Kite Demography: Annual Report 2002*. Report prepared for US Fish and Wildlife Service. 2002.

I used quotes and images from several published works.

- Chapter 1: The map in figure 1.3 was adapted with permission from P. W. Sykes Jr., "The Range of the Snail Kite and Its History." *Bulletin of the Florida State Museum: Biological Sciences* 29 (1984): 211.
- Chapter 2: The quote starting with, "We were rounding the corner, and we both looked," was spoken by Dr. Caroline Poli in a video by Justin Bright of WUFT News, under the title: *The Future of Florida's Endangered Snail Kite Is Unclear*, available on YouTube; used with permission. The quote about the snail kite's call sounding like "a wooden dowel turning in a tight hole" comes from Dr. James Rodgers Jr., "Diary of an Everglade Kite Survey," *Florida Naturalist* 56, no. 3 (1983): 8.
- Chapter 3: The map in figure 3.1 was adapted with permission from S. E. Haas, J. Martin, W. M. Kitchens, and R. T. Kimball, "Genetic Divergence Among Snail Kite Subspecies: Implications for the Conservation of the Endangered Florida Snail Kite *Rostrhamus sociabilis*," *Ibis* 151 (2009): 181–85. (© 2009 John Wiley & Sons, Inc.)
- Chapter 7: The quote starting with, "This scenario—cyclic drought, a falling water table," is from J. E. Takekawa and S. R. Beissinger, "Cyclic Drought, Dispersal, and the Conservation of the Snail Kite in Florida: Lessons in Critical Habitat," *Conservation Biology* 3 (3) (1989): 302–11 (© 1989 John Wiley & Sons, Inc.). The map in figure 7.2 is adapted with permission from figure 1 of the same article. The

suggestion that snail kites had disappeared into a black hole in 1981 was made by Dr. James Rodgers Jr., on page 6 of his *Florida Naturalist* article referenced above.

- Chapter 7: Figure 7.1 and the quote starting with, "You feel like sitting down," originally appeared in an article by Marion Phelps that was published in *Fort Lauderdale Sun-Sentinel*/TCA; used with permission.
- Chapter 9: The quote beginning, "Oh Sam, I DO know the secret!" is from *Sam the Snail Kite and the Secret of the Everglades* published by the City of West Palm Beach in 2025, written by Vera de Chalambert and illustrated by Christina Eure (available for purchase through the Grassy Waters Preserve website). The poem "Hope Is the Thing with Feathers" by Emily Dickinson is from *The Poems of Emily Dickinson,* edited by Thomas H. Johnson (Belknap Press of Harvard University Press), Copyright © 1951, 1955, 1979, 1983 by the President and Fellows of Harvard College.
- Chapter 10: The quote starting with, "Because Florida snail kites are considered," is from page 66 of National Research Council, Division on Earth, Life Studies, Board on Environmental Studies, Water Science, Technology Board, and Committee on Independent Scientific Review of Everglades Restoration Progress (CISRERP), *Progress Toward Restoring the Everglades: The Fourth Biennial Review, 2012* (National Academies Press, 2012).
- Chapter 15: The quote starting with, "The federally endangered bird, the snail kite, was faced with," is from a 2017 article by Beverly James titled "UF Study: Bird Evolves Virtually Overnight to Keep up with Invasive Prey" from the University of Florida News website; used with permission.
- Chapter 23: The quote from Dr. Paul Sykes Jr. about "selected 'islands' of [snail kite] habitat" appeared in his article "Snail Kite Use of the Freshwater Marshes of South Florida," *Florida Field Naturalist* 11, no. 4 (1983): 2.
- Chapter 25: The quotes starting with, "Continued IOP operations" and "However, because snail kites are long-lived," are from US Fish and Wildlife Service, *Biological Opinion on the Continuation of the Interim Operational Plan for the Protection of the Cape Sable Seaside Sparrow,*

November 17, 2006 (Service Consultation Code: 41420-2007-F-0045), South Florida Ecological Services Office.

- Book epigraph, and also in chapter 27: The quote, "Systems of people and nature co-evolve in an adaptive dance," appeared in L. H. Gunderson, "Adaptive Dancing: Interactions Between Social Resilience and Ecological Crises," in *Navigating Social-Ecological Systems: Building Resilience for Complexity and Change* (2003): 33–52. He gave permission to use that quote, and noted that he found the term "adaptive dance" in Carl J. Walters, *Adaptive Management of Renewable Resources* (Macmillan, 1986).

# INDEX

Page numbers in *italics* refer to illustrations.

HILARY FLOWER is associate professor of environmental studies at Eckerd College, where she teaches about wetlands and global environmental change. She holds a PhD in ecohydrology from the School of Geosciences at the University of South Florida, Tampa. Her peer-reviewed research focuses on how climate change and sea level rise affect the Everglades. She enjoys amateur bird-watching and is an enthusiastic naturalist interested in how humans and wildlife interact.